Psyched on Bikes
The bicycle owner's handbook

Psyched on Bikes

The bicycle owner's handbook

B. ANDREW RENTON

TAB BOOKS
Blue Ridge Summit, PA

FIRST EDITION
FIRST PRINTING

© 1992 by **B. Andrew Renton**.
Published by TAB Books
TAB Books is a division of McGraw-Hill, Inc.

Printed in the United States of America. All rights reserved. The publisher takes no responsibility for the use of any of the materials or methods described in this book, nor for the products thereof.

Library of Congress Cataloging-in-Publication Data

Renton, B. Andrew.
   Psyched on bikes : owner's handbook / by B. Andrew Renton.
     p.   cm.
   Includes bibliographical references and index.
   ISBN 0-8306-2021-4   ISBN 0-8306-1987-9 pbk.
   1. Bicycles.   I. Title.
TL410.R36   1992
629.227'2 dc20   91-46793
                                            CIP

Acquisitions Editor: Kimberly Tabor
Book Editor: Robert Amsel
Designer: Jaclyn J. Boone
Director of Production: Katherine G. Brown
Cover Design: Holberg Design, York PA.
Cover Photo: Susan Riley, Harrisonburg, VA.   ATS

Contents

Preface   ix

1   So, you are buying a bike   1
     Cost   6
     Summary   7

2   Frames, wheels, & tires   9
     Design features of frames   9
     Multiseat machines   11
     Tricycle configuration   12
     Frame materials   14
     Wheels   16
     Tires   21
     Fenders   27
     Summary   28

3   Saddles & steering mechanisms   29
     Saddles or seats   29
     Cycle-Shock: shock-absorbing seat post   35
     Steering mechanisms   37
     Summary   45

4   Transmission systems   47
     Pedal mechanisms   49
     Friction   53
     Lubrication & other maintenance   55
     Transmissions   56
     Summary   64

5 Braking mechanisms 67
A caliper brake hybrid   67
Coaster brakes   68
Caliper brakes   69
Brake levers   72
Disk brakes   74
Drum brakes   74
Summary   75

6 Accessories 77
Helmets   77
Lights   78
Bicycle computers   84
Rearview mirrors   86
Reflectors   87
Locking devices   89
Bike carriers   91
Portable working & parking bike stand   93
Clothing   93
Personal waist & backpacks   94
Water containers & cages   94
Toe clips & straps   95
Tools   95
Lubricants   101
Chain conditioners & cleaners   103
Summary   104

7 Safe driving techniques 107
Carriers   107
Training wheels   108
Kids & safety rules   111
Insurance   113
Municipal ordinances   114
State (USA) & provincial (Canada) regulations   114
Bicycle paths   115
Bicycle tours & competitions   116
Practical driving skills   116
Summary   120

8 High-tech developments 123
Types of frames 124
Types of wheels 127
Lights 130
Electronics 131
Tires 131
Summary 133

9 All-terrain & mountain bikes 135
Cost 135
Manufacturers 136
Summary 144

Manufacturers & products 145

Glossary 155

Index 165

Preface

Initially, I became psyched on bikes when I was 13 years old. Al Potter—my Ottawa boyhood pal—had a three-speed Raleigh with a Sturmey-Archer internal shifting mechanism. Al loaned me his bike frequently, and I would roar off to the Gatineau Hills, northeast of Ottawa, in the beautiful province of Quebec. About two hours later I would return to find Al having a fit. Had Renton had an accident? Certainly not! I was just psyched on *his* bike.

Eventually, by saving money as an *Ottawa Journal* newspaper carrier, I purchased my own Raleigh. I had no fancy derailleur then—just a one-speed coaster brake bike.

World War II came about. I served as a Flying Officer, flight engineer aircrew type, in the Royal Canadian Air Force. After the war, I studied mechanical, electrical, and electronic drafting. I became a secondary school drafting teacher, then a departmental head, then a technical education consultant, and finally, the principal of a special education junior vocational high school. All these career steps helped prepare me for my latest challenge—as author and illustrator of *Psyched on Bikes*.

I also organized and produced a series of Mobile Bicycle Repair Programs for many of the elementary schools under the jurisdiction of the Ottawa Collegiate Institute Board, now the Ottawa Board of Education.

Following this venture into bikes and biking, I was contracted by a Toronto publisher to write and illustrate a bike text. Unfortunately, this project was never completed, but the dream of writing such a book lived on.

About three years ago, my wife Joyce—an invaluable typist and preliminary screening agent—purchased as a gift for me, an excellent TAB book on the U.S. Space Shuttle program, up to the Challenger disaster. I was wintering in Florida during that sad time.

At any event, a casual letter addressed to TAB's president started the bicycle wheel rolling again. After many letters to vice presidents and to Kim Tabor, the Senior Acquisitions Editor, TAB and I committed ourselves. I owe Ms. Tabor and all the personnel at TAB a special thanks.

I must mention also the kind assistance of the many American, Canadian, Italian, and Japanese bicycle and bike components manufacturing firms, who provided me with the latest technical information and photo illustrations.

Special thanks are also due to Janice Klassen, who turned my technical "roughs" into finely tuned executed technical illustrations.

I also appreciate the help extended to me by Caron and Dave Garratt, the owner-operators of franchised branches of Cyclepath, Ottawa, Canada; Harry Neyhart of the Village Bike Shop in Hobe Sound, Florida; and Peter Haggerty of the Bike Stop in Canada's capital city.

CHAPTER ONE

So, you are buying a bike

Bicycling has been enormously popular for the past decade, even longer. The older generation was satisfied with a single-speed, coaster brake machine. A snappier model with the Sturmey-Archer, three-speed internal shift was considered a big deal. Although dads and moms still purchase single-speed models for their pre-teenage kids, they are likely to buy derailleurs with caliper brakes for more style-conscious adolescents.

Biking is carried out for different reasons. The pre-teenager sees it as fun—something other members of the "gang" do. The 15- or 16-year-old regards biking as a means of transportation. However, motorized scooters, mini-motorcycles, and automobiles (where the legal driving age is 16) often seduce teenagers away from bicycles.

But, to younger men and women who are "psyched on bikes," biking is considered healthy. It improves one's physical condition and keeps body muscles toned up, especially legs, thighs, and buttocks. Additionally, bikers have a unique sense of involvement. There are biking clubs and associations to belong to, and cross-country and interstate trips to make. During a trip to the People's Republic of China, I saw bicycles being disgorged from a Boeing 747 at Beijing International Airport. I soon discovered that an American group was planning a bicycle tour south to Shanghai, a 650-mile journey. When you are psyched on bikes, you soon learn that the world can be your oyster.

What about the unique clothing, footwear, and other accessories that bikers must have? These include proper shoes, white socks, silk-screened tee shirts, approved safety helmets, gloves, and colorful skin-tight spandex bicycle shorts. Not any old shorts will do, but designer types of the right material and correct cut.

Bicycling is not kid stuff anymore! Young business people, and some not-so-young, have taken to this sport in a big way. It has become a big-league industry with high financial stakes.

Many different bicycle designs and styles are fabricated by American, European, and Asian manufacturers. Some firms specialize in the lower-priced single-speed, coaster brake types (Fig. 1-1); others, in the middle-range-priced three-speed, internal shift styles (Fig. 1-2). In Italy, France, and Japan, very expensive, custom-designed racing bikes are designed and manufactured (Fig. 1-3). A small number of

Fig. 1-1. Typical single-speed coaster-brake bike.

companies produce three-wheeled work bicycles (Fig. 1-4). These machines have heavy-duty frames, which include metal containers for transporting items. High speed is certainly not a feature of their design; strength and dependability are.

So, you are buying a bike   3

Fig. 1-2. Typical three-speed internal-shift bike. This model is the Collegiate, manufactured by Schwinn.

Fig. 1-3. Typical, multispeed derailleur bike. This model is the Schwinn LeTour.

Additionally, off-road and mountain bikes are now available (Figs. 1-5 and 1-6).

How much should you pay for a bicycle? The sky is the limit. An overstatement? Custom-designed bikes for a professional racer in the Tour de France or other big-name events have spiraling price tags well

Fig. 1-4. Typical three-wheeled utility bike made by Worksman.

Fig. 1-5. Typical ATB mountain bike. Shown here is the Montana Sport, made by KHS.

Fig. 1-6. Typical ATB mountain bike shown is the Montana Crest, made by KHS.

beyond a thousand dollars. The materials are exotic and expensive, including special thin-walled, high-tensile steel. Components manufactured from lightweight materials containing immensely strong graphite filaments are also used. Each bike frame is made-to-measure for a particular individual, which greatly increases the cost. Yes, such bikes are handmade with loving, meticulous care.

When welding is done, the welds are works of art. Bearings, gearing and shifting items, braking pads, cables and levers—all are of advanced technological design and are manufactured from the finest materials. The price tags on such machines are quite astounding. Are these bikes insured for theft and road hazards? You bet they are!

Table 1-1 gives some insight into the possible costs of various types. Please note that the cost figures are only approximate. Place, time of purchase, and other variables will have a distinct bearing on a particular unit's cost.

In addition, a bike's usage will have a bearing on the projected or anticipated cost. Is it a gift for a youngster or an adult? What is the buyer's financial situation? Certainly a young professional without any family responsibilities will look differently at bike costs than would a parent or grandparent. Since bicycles have widely ranging price tags, purchasers should pay what they can afford.

Table 1-1 Costing chart.
(Figures are approximate)

General Classification	Approximate Price
ATB	$160 to 500
Coaster	100 to 200
City	300 or more
Hybrid	180 to 1000
MTB	300 to 2000 plus
Recumbent	1500 to 3000
Road	300 to 5500 plus
Tandem	600 to 1500
Three-speed	150 to 250
Tricycle (3-speed)	400 to 600
Work (Industrial) Bike	200 to 400
Work (Industrial) Trike	400 to 800

Legally, a bicycle is a non-motorized vehicle. But in many respects, the operator of this human-powered machine must drive it as if it were motorized. A responsible driver should obey stop signs and stoplights, should drive on the same side of the road as motor vehicles, should respect the rights of pedestrians (mobile or stationary), and should recognize the rights of motor vehicles (cars, trucks, and motorcycles).

Horseplay and general foolhardy bicycle driving is tolerated much less by today's law enforcement officers than such behavior was two or three decades back. Too many young and not-so-young bicyclists are receiving serious, sometimes fatal, head injuries on the highways and byways of North America.

Drive wisely, courteously, safely, and live to be psyched on bikes for many, many years to come.

Cost

The following factors might have a bearing on the cost of a particular bicycle, tandem, or tricycle. Price ranges may be as low as $99 to $150 or as high as $5500 and above for a custom-built road racer.

Quality U.S. and imported European machines are more expensive in Canada than in the U.S. because Canadian dollars are worth less than U.S. ones.

The bike retailer's location might influence pricing. Is it located in a high- or low-rent area? Is the business a single proprietorship? Are sales volumes high enough to justify special sale prices from time to time? Geographical location might also play a part. In many northern U.S. states and much of Canada, the retailer slashes prices as winter approaches.

One major sporting goods entrepreneur in Ottawa, Canada is currently selling bikes at extremely low prices and "throwing in" a free pair of cross-country skis. The seller might be unloading last year's stock, but the frugal buyer is getting a terrific bargain. In conclusion, bike prices are not etched in stone, and a potential buyer should shop around.

Summary

- Economical transportation, pleasure, and exercise are all valid reasons for biking.
- Clothing and other accessories are available to fashion-conscious bikers.
- Bikes vary in price to fit the individual's needs.

CHAPTER TWO

Frames, wheels, & tires

A major bicycle component is the frame. It may look flimsy, but it is very strong structurally. Strong, but not invincible. A bicycle frame can be twisted out of front-to-rear vertical alignment. Worse, the main horizontal top tube and diagonal down tube can jackknife when the driver strikes an immovable object head-on, or is involved in a tangle with a moving car or truck. At this stage, a damaged frame is the least of the bicyclist's problems.

Design Features of Frames

Figure 2-1 is representative of the configuration of many, if not most, single-seat (or single-passenger) bicycles. The point at which the front fork swivels within the steering head could be prone to minor damage in an accident. However, the steering head, the down tube, and the horizontal top tube are secured to it by sizable weldments. Excessive force exerted against the front wheel is thereby transferred to the front fork, causing the fork to jacknife slightly or the wheel rim to bend out of vertical alignment. Usually the wheel rim twists. If the damage is slight, the rim should be re-trued by adjusting the spokes.

A spoke wrench, shown in Fig. 2-2, is inexpensive and easy to use with training. Its use by an unskilled operator may further increase rim misalignment.

Fig. 2-1. Frame outline of a single passenger bicycle.

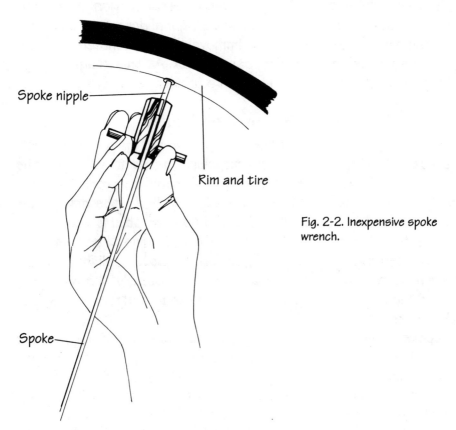

Fig. 2-2. Inexpensive spoke wrench.

Note the bike frame's triangular construction. It is similar to the triangular structures found in bridge construction. When you consider the abnormal stress placed on a bicycle frame by youngsters who intentionally jump concrete curbs for the fun of it, you can see that bicycle frames are exceedingly robust. This strength is further enhanced by the design of the tubing used, the material from which the tubing is formed, and the quality of the attaching weldments. Before enlarging on these aspects of frame strength, I shall describe the structure of multiseat and three-wheeled machines or tricycles.

Multiseat Machines

The geometry of a two-seater machine is shown in Fig. 2-3. This particular model is a Trailmate tandem Easy Ride. Note that the top tube is angled downwards toward the rear seat. A secondary tube is welded at four places terminating at the rear of the chain stay.

Because the bike has two driving positions, the down tube from the forward steering bars is attached by weldments at the steering head, at the intersections of the first down tube and the first seat tube, at the second seat tube, and at the second sprocket-axle tube.

Fig. 2-3. Trailmate's tandem Easy Ride bicycle.

Finally, at the rear wheel chain stay, all six structural tubes are neatly angled and welded. This design, which uses high-tensile steel, offers great strength.

For racing buffs, the Schwinn Bicycle Company of Chicago, Illinois, manufactures a very slick model known in the trade as the Schwinn Duo Sport. As shown in Fig. 2-4, this excellent machine has racing type seats and handlebars, toeclips, and a multispeed, external gear transmission system. Note that there is no top tube between primary and secondary seat tubes, as in the Trailmate Easy Ride model. Also, the Trailmate Tandem Classic does not have top tubes. Either seat position may be used by a driver wearing a skirt. Trailmate speaks of its Tandem Classic in endearing, romantic terms.

Fig. 2-4. Schwinn's Duo Sport bicycle.

Tricycle Configuration

The particular model referred to is not a heavy-duty delivery tricycle, but one frequently used by senior citizens in retirement communities. Figure 2-5 is a photograph of one manufactured by Trailmate of Sarasota, Florida, that goes by the product name EZ Roll Regal. Notice that the down tube is additionally reinforced by a smaller diameter tube welded to the bottom of the steering head and to a mid-bottom point of the down tube. This smaller tube is also secured to the horizontal member of the rectangular cross section, located somewhat above the axis of the drive sprocket. The comfortable seat has a three-position anchor design. The dual rear wheels provide

Fig. 2-5. Trailmate's EZ Roll Regal tricycle.

great stability and operating ease. The overall geometry of the EZ Roll Regal's design—similar in other three-wheelers—offers excellent stability while the bike is in motion. This tricycle is not a high-speed machine, but one that offers comfort, safety, and security—ideal for riders advanced in years.

Work Tricycles

Worksman Trading Corporation of Ozone Park, New York, manufactures heavy-duty work or industrial bicycles, as well as a wide range of adaptable tricycles and front-load designs. Figures 2-6 and 2-7 show Worksman models.

These specialized tricycles have a payload of 250 pounds, which is 3½ times the amount that a conventional adult tricycle can transport. Sprockets, spokes, and chains are of a heavy-duty design and manufacture. Wheels are semipneumatic, eliminating the messy problem of tube repair or replacement.

New design features have been introduced wherever rough, heavy, and continuous usage could result in failure. The company was founded by Morris Worksman, whose designs have been popular with workers throughout the century.

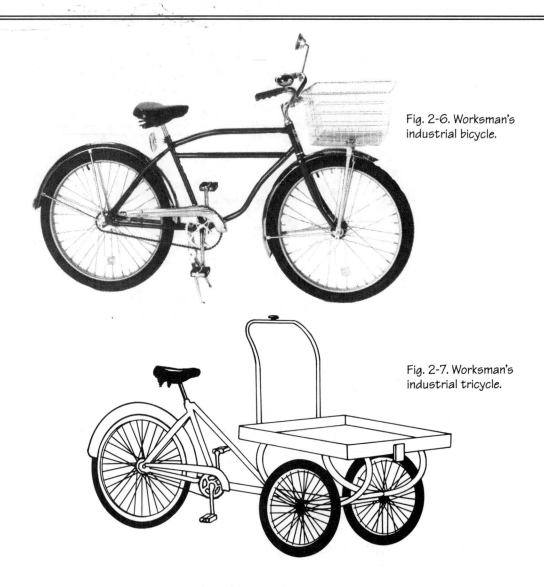

Fig. 2-6. Worksman's industrial bicycle.

Fig. 2-7. Worksman's industrial tricycle.

Frame Materials

The material in the frame is steel, which is basically iron containing a low content of carbon hence, the name, carbon steel. To increase its strength, anticorrosive characteristics, and other qualities, the carbon-steel is alloyed with small quantities of other elements.

In general, the steel in inexpensive bikes ($100 to $200) falls far short of the chrome-moly (molybendum) high-tensile steels used in better quality machines. In the better bikes, the wall thickness of the chrome-moly alloy tubing is reduced because of the tube's inherent strength. The bicycle's overall weight is therefore considerably reduced, a major design factor for expensive racing bicycles, such as those used in the Tour de France.

Less expensive bikes still use alloy steel tubing, but it is not as strong as chrome-moly tubing. Nor is it as costly to produce. If you are not racing for fame, award certificates, plaques, trophies, or money, you might not require a chrome-moly steel tube frame bike.

Fig. 2-8. Cannondale's SR-400 bicycle has a frame weight of only 3.45 pounds.

Lightweight aluminum alloys with a hard, anodized finish are used extensively in bicycle components. While not having the tensile strength of steel alloys, their light weight and resistance to corrosion are important. Cannondale of Georgetown, Connecticut, builds a road racer, the Cannondale SR-400 (Fig. 2-8, above), that utilizes oversized aluminum alloy tubing to eliminate frame flex.

Because of Cannondale's experience with using aluminum alloy tubing in bikes, it has developed expertise and skill with this substance. For example, in its Tandem bicycle, Cannondale uses oversized aluminum-alloy tubing to greatly increase stiffness, provide more room for the stoker (rear rider), and reduce the frame's weight by 25 percent over a chrome-moly model. The front wheel fork, however, is still manufactured from chrome-moly steel for obvious design reasons—namely, high strength.

Tube configuration can add to the rigidity and strength of a member. Most tubing is elliptical in cross section. Different manufacturers often use variations of standard geometric shapes to advantage. It is unlikely for manufacturers to produce every component from start to finish in their assembly shops. For decades, numerous companies have been involved in bicycle components specialization and "raw" materials manufacturing as they have been in the automotive and the electronic industries.

Titanium steel tubing is found in the very expensive frames of some mountain bikes. Its weight-to-strength ratio makes it the best of steel alloys. However, its high cost precludes its use in great amounts. For fighter aircraft and space vehicle components involving government spending, there appears to be no financial limit in the use of this exotic material. However, I recently purchased a set of 13 Vermont-American titanium steel drill bits, from $1/32$-inch diameter to $1/4$-inch diameter, for less than $20. Perhaps the day of titanium steel products for the humble citizen has arrived.

Wheels

Rims

The rim's function is to contain the tube and provide a means of securing the tire to it. As a consequence, the rim is manufactured from a steel alloy, chromium-plated for both appearance and rust prevention. Stainless steel is also used for rim construction.

An aluminum alloy, Number 6061-T6, manufactured by the Sun Metal Products of Warsaw, Indiana, has recently invaded the rim design arena. It has an ultra-hard anodized finish. Also, Alesa Manufacturing Company, a major rim manufacturer throughout the world, now produces a fiberglass-disk racing rim. Its highly specialized design eliminates the need for spokes. Todson, Inc., of Deer Park, New York, is the sole U.S. distributor of this unique Alesa product.

Not to be outdone by Alesa, Sugino of Japan now produces an extremely lightweight, high-strength wheel, which incorporates tensioned Kevlar strings, sandwich-bonded between two plastic sheets.

Frames, wheels, & tires   17

In addition to space-age materials found in rim and wheel manufacture, guaranteed puncture-proof tubes have invaded the bike market. With such fantastic products, it's easier to become psyched on bikes.

When a rim is severely damaged in an accident, it should be replaced. Minor damage that produces wheel wobble or misalignment probably can be corrected by having the wheel trued by a competent technician. If the bike is inexpensive, a person with reasonable technical skills may be able to assess which spokes need tightening or loosening. Unfortunately, you need a special clamp to hold the wheel axle during this procedure, and a gauge to indicate how far the rim is off "true." A professional model truing fixture is shown in Fig. 2-9.

Fig. 2-9. Park Tool's professional model wheel-truing tools.

I have used—as have other bike enthusiasts—an inexpensive spoke wrench, Fig. 2-2, to carry out minor rim truing on inexpensive coaster-brake models. Don't be chicken! Give it a whirl! You can always take the wheel or bike to an authorized service technician afterwards, if required.

The rim provides the anchoring points for the spokes that emanate from either the front or rear wheel hubs. Since the bike's operation depends upon the trueness and rigidity of the rim in conjunction with the tube and tire, its manufacture precludes relative lightness and high strength. These qualities are attained when the manufacturer forms the rim with dies that transform flat, steel stock into a shape that provides sufficient tire anchoring when applied.

Figure 2-10 depicts the cross-sectional shape of a variety of wheel rims and shows how the tire is held in close contact with the rim under air pressure from the tube.

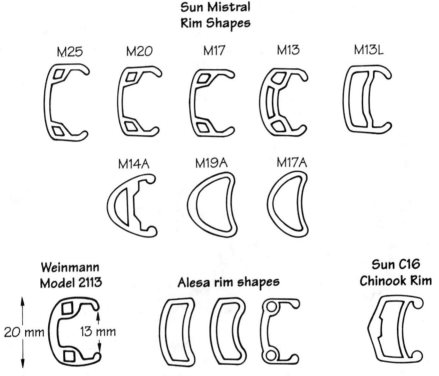

Fig. 2-10. Cross-sectional shapes or rims.

Figures 2-11 and 2-12 show the advanced designs of Alesa's fiberglass racing discs and Sugino's Kevlar tension-stringed model.

Spokes

Although an alloy steel or aluminum alloy rim has a certain amount of stiffness because of its design, it is only part of the assembled wheel. Without the proper attachment of spokes to the front and rear hubs and to the rims, the wheel would be useless.

Frames, wheels, & tires   19

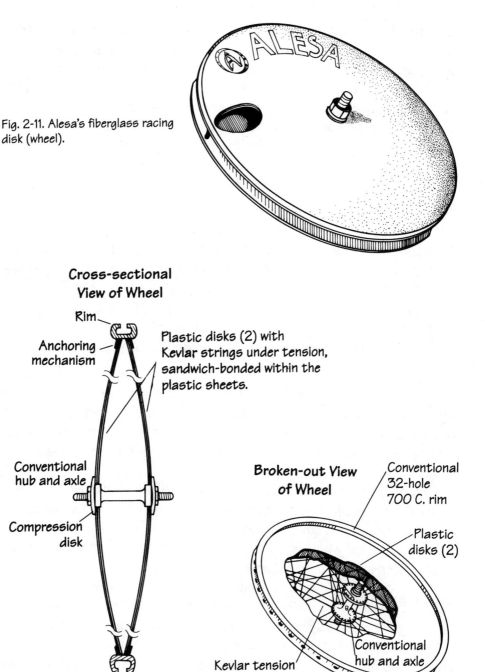

Fig. 2-11. Alesa's fiberglass racing disk (wheel).

Fig. 2-12. Sugino's Kevlar tension-stringed wheel.

Thus, the spokes are crucial components in the overall assembly. Because of their positional geometry, the spokes are always in tension: permanently secured to the hub flange and threaded into a simple attachment device, a nipple located in the rim's center. A composite, fabric-rubber tape fits snugly against the inner periphery of the rim to provide protection against tube abrasion under pneumatic pressure. Otherwise, chafing would occur followed by eventual air pressure loss.

The number of spokes per wheel seems to be a design variable, dependent upon the manufacturer. The following information tabulates a few models:

Sturmey-Archer rear-wheel hub—15 each side, for a total of 30.
Shimano "D" type hub—11 each side, for a total of 22 (coaster-brake).
Majestic coaster-brake hub—19 each side, for a total of 38.
Bayliss-Wilby front hub—19 each side, for a total of 38.
Shimano 5 sprocket—19 each side, derailleur rear hub, for a total of 38.

Other variations are possible, but regardless, the function of the spoke is constant: it is a small diameter, linear, metal-connecting device that attaches the wheel hub to the rim. It offers low wind resistance, is light in weight, and above all, is simple to install and economical to manufacture.

Most spokes are made from 24 ga (2.0-mm diameter) galvanized steel. Better quality ones, manufactured from #14 ga stainless steel, are also available. Lighter-weight spokes with a diameter of 1.8 mm (15 ga) are used when reliability is not of prime importance; the rider is a lighter-weight person, and the manufacturer is concerned with reducing overall bike weight. The manufacturer can reduce weight by 40 to 50 grams per wheel by using 15 ga spokes. Straight-gauge spokes are the most popular because they are economical to produce. However, double-butted ones with thinner center sections, such as those manufactured by Wheelsmith, offer the same strength as identical straight-gauge spokes, but have less weight.

Wheelsmith also produces an aerodynamically shaped spoke, the ACE (Aerodynamic Competitive Edge), to overcome wind resistance.

The hub holes in the wheel rim must be slotted in order to use the ACE spoke. These spokes are available in either the 1.8 or 2.0 mm configuration.

Spoke nipples, Fig. 2-13, are made of Duristan (a Wheelsmith finished brass material) or in an anodized, extra-hard aluminum alloy. The nipple's function is to secure the spoke to the wheel rim.

Spokes that are neither galvanized (hot-zinc dipped), nor formed

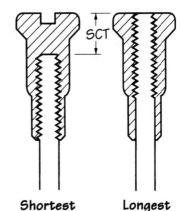

Fig. 2-13. Wheelsmith's nipple details.

from stainless steel, are available in 2.0-mm diameter plain steel. I would hesitate to recommend plain steel spokes because of the possible rust problem.

Tires

Because a considerable variety of bicycle designs are available—touring, road-racing, off-road or mountain driving, goods delivery—the type of wheel, inclusive of its tube and tire, is anything but a constant.

Touring bikes utilize a tire width and tread design consistent with reasonable comfort.

Road-racing bicycles have a narrow tire profile and a very minimal (if any) tread. High speed is the major design criterion.

Off-road and mountain bike tires have a wide configuration and a knobby tread. Sometimes they are formed from Kevlar, or at least with a Kevlar rim bead, for additional strength and a maximum grip of the rough terrain in contact.

Delivery bikes or tricycles have wheels with wide tires so that the considerable weight being supported by the bike—up to 250 pounds of freight, the operator's weight and the tricycle's weight—is sufficiently supported. Treads might be deeper than those used in a touring bicycle.

The air pressure to be maintained on the above machines also varies. Table 2-1 offers invaluable tire inflation information.

Table 2-1 Approximate tube air pressure for different bikes

Size of Tire	Rec. Air Pressure in lbs per sq. in. (psi)	
12"x1⅜"	30–40	
16"x1⅜"	30–40	
18"x1⅜"	35–45	
20"x1⅜"	45–50	
24"x2⅛"	35–45	
26"x1¼"	45–50	
26"x1⅜"	45–50	
26"x1¾"	30–35	
26"x2⅛"	35–45	
27"x1¼"	85–100	
700 cmx1"	85–100	
700 cmx1⅛"	85–100	
———Tubulars—27-inch———		
Road Condition	Front Wheel	Rear Wheel
Uneven surface	90 psi	90–100 psi
Very smooth surface	90–120 psi	100–140 psi
Road Racing Bikes	90 psi (depends on road conditions)	90–100 psi
Touring Bikes	85–90 psi (depends on load/road conditions)	85–100 psi

If you are a heavy person, inflate the tire to about 5 psi higher than shown in Table 2-1. Should the tires still bulge markedly, increase the pressure somewhat. When your bike is used by yourself, only a small amount of bulging should be tolerated.

When you subject a tire to hard usage, such as your quickly jamming on the brake of a coaster-brake design bike, you will wear down the tread at an accelerated rate, even on a new model. Jumping curbs and doing "wheelies" can create so-called "stone" bruises on the tires' running surface. Not maintaining sufficient tube air pressure will result in fractures to the tire wall (linear crack lines) after a while. The variable weather conditions in your area and the place where you store your bike might have a deteriorating effect on the bike's overall condition, including its tubes and tires. Bicycle design and manufacture, while good or even excellent, cannot eliminate the effects of poor treatment or neglect.

Semi-pneumatic Tires

A tubeless tire manufactured by PolyAir Tires, Inc., of Calgary, Alberta, Canada, is the answer to a biker's prayer. Hundreds of thousands of microscopic air cells are trapped in its matrix of incredibly tough polyurethane. It is as tough as nails and as smooth as air. The PolyAir MCP Tire is even guaranteed for a full year. Figures 2-14 and 2-15 illustrate the nature of this hybrid tire and the correct installation procedure to the wheel rim.

Enlarged, partial cross-sectional view of tire, showing nail impaled in a matrix of polyurethane and hundreds of thousands of microscopic air cells.

Fig. 2-14. Tubeless, PolyAir, hybrid tire. PolyAir Tire, Inc.

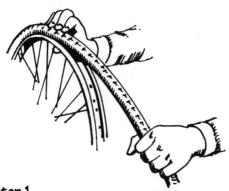
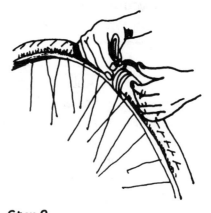

Step 1
Place the tire on the rim and fit one section of the tire into the rim channel so that the tire shoulder is flush with the rim edge.

Step 2
Tie this section into place with a wire or heavy string wrapped around tire several times so that the tire will not pop out of the rim channel.

Step 3
Take a prying instrument and starting at the point where the tire is fastened, lever the tire onto the rim. Move the instrument 3 or 4 inches and lever again. Repeat this procedure until installation is complete. The tire will stretch over the rim and fits snuggly. Don't be afraid to pry. When the tire is on the rim, remove the instrument and the wire or string.

Step 4
Make sure the tire shoulder is flush with the rim all the way around. If it is not seated properly, gently bounce the wheel on the ground at the misfitting point until the tire fits evenly.

To remove:
Take pliers or vice grips and place jaws over tire, not touching the rim, and pull to one side exposing the inside of the tire. Insert the prying instrument between the tire and rim and pry the tire off.

Fig. 2-15. Installation procedure for PolyAir tires.

Tubes

A tube must have sufficient air pressure to maintain both the rider's weight and the bike's weight. Make sure the tube size is commensurate with the tire size. When you install a new tube, or simply replace a repaired one, make sure that the tube's air stem is properly placed relative to the hole in the rim: the stem should angle inwards toward the wheel center and form a 90-degree angle with the rim. This placement is illustrated in Figs. 2-16 and 2-17.

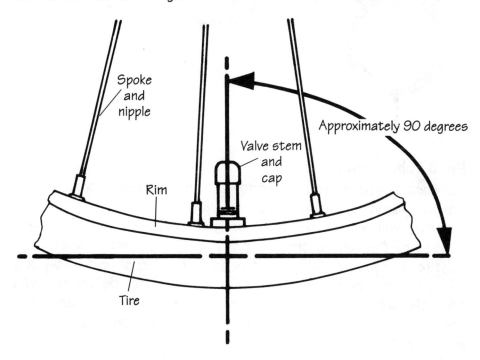

Fig. 2-16. Correct positioning of tube and air valve stem when viewing wheel from side.

Tube prices vary. The least expensive rubber tube might not be the most economical in the long run. Many cheaper tubes do not have a threaded stem nor a threaded locking washer. They depend entirely on the friction created between the tapered rubber stem and the hole in the rim it penetrates to maintain accurate positioning. I regard this as a poor design feature, however economical.

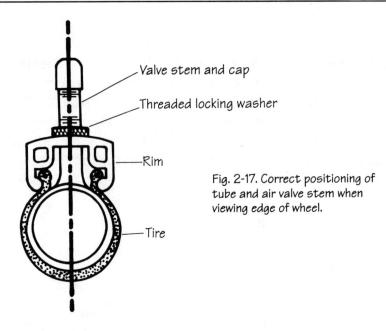

Fig. 2-17. Correct positioning of tube and air valve stem when viewing edge of wheel.

Tire & Tube Removal

The wheel assembly must be removed from a bike prior to changing a tire or replacing or patching a damaged tube.

After you remove the front or rear wheel, deflate the tube by depressing the valve stem core pin. Using simple lifting levers, force the tire bead progressively over the rim. A number of reasonably large, plain screwdrivers can do the job if you are careful. Be careful not to sever the tube by pinching it between the inside edge of the rim and the screwdriver blade. Soon, the tire will be freed from one edge of the wheel rim.

Unless the tire is damaged with wall splits or abrasions, or has bald areas without tread, you needn't remove it completely from the wheel.

Pump some air back into the deflated tube so that any hole can be spotted. Then draw a white chalk mark around the hole. Deflate the tube once again. Gather the tube into a large "mass." Depress the valve-stem core pin, while squeezing the so-called "mass". This action will eliminate enough residual air in the tube to allow you to clean and "rough up" the leak area with the proper tool. Remaco, Inc., of Northvale, N.J., manufactures a repair kit that contains the necessary cleaner/buffer to remove foreign materials that might prevent a proper bond between the patch and the tube. The kit also includes vulcanizing fluid (patching cement), feathered patches, and emery paper. Refer to Fig. 6-28.

A feathered rubber patch is ideal for repairing a nail hole, since the patch's outer edge does not create any noticeable "bump" problems. It also bonds smoothly to the tube when pressed or rolled in place. Patching kits invariably have simple directions printed on their containing boxes. Read and follow these directions.

Tube replacement within the tire is a simple procedure. Make sure that the tube lies flat in the tire. The valve stem should be centrally located within the cross section of the tire. Make sure that when the tube/tire combination is offered to the rim, the valve stem penetrates its opening properly. Work this combination onto the rim, making sure that no tube part is outside the tire. Once again, beware of pinching the tube. Hurrah! The tube is home safely once more. Apply some air pressure to the Schraeder valve and check again to see if you've done your work correctly. Then pump up the tube to the recommended air pressure. Reinstall the wheel to its appropriate fork. Check your work again. You should be ready to ride.

Fenders

Not much can be said about fenders. Some bikes have them; others do not.

Road racers do not have them since light weight and high speed are essential.

All-terrain bikes (ATBs) and mountain bikes avoid them because of rugged trail conditions. Mud and other natural debris could jam between the outer tire and inner fender surfaces.

Touring bikes often have fenders because comfort, not high speed, is essential. Also, people who ride touring bikes do not wish to get their clothing soiled.

Fenders are made from lightweight plastics, aluminum alloy, or steel. Steel or aluminum fenders can be reshaped if bent out of their intended configuration. At each fender attachment location, you will notice a machine screw in a threaded hole, held from backing off by a lock washer under the screw head. Occasionally the screws become loose or fall out, causing the fender to vibrate. Occasional spot checks can eliminate this minor problem.

Summary

- **Frames**   The overall geometrical form, the materials and cross-sectional tubing shape in conjunction with excellent weldments, produces an exceedingly strong, rigid, and lightweight frame. Chrome-moly, high-tensile steel is most commonly used in high-quality bikes. Large, cross-sectional shaped aluminum alloy tubing is used in lightweight models without sacrificing strength and rigidity.
- **Wheels**   Wheels are constructed from steel alloy and then chromed. However, fiberglass and Kevlar plastics have now invaded the standard spoked-wheel configuration.
- **Tires**   A cross-sectional shape, tread design, and tube air pressure are required and vary according to the bike's designated usage.
- **Tubes**   Quality and price may vary. The least expensive is not necessarily the best buy. The use of approved patch kits for hole repairs is well within the technical expertise of teenage and adult bicyclists.
- **Fenders**   Certain bikes do not use them, in order to minimize weight and maximize speed. In inclement weather, fenders can largely eliminate the mud-spattering of clothing.

CHAPTER THREE

Saddles & steering mechanisms

Because of human anatomy, most seats or saddles are roughly triangular. Although I shall use "seat" or "saddle" interchangeably, some three-wheelers and all recumbent bikes have "seats," as that term pertains to a chair, even if modified by design or materials.

The specific design function of every bicycle or tricycle will determine the seat's shape and materials used, and whether or not the seat will have shock-mounted characteristics.

Saddles or Seats

The triangular seat portion is an individual component attached to a steel bracket device, which allows the seat to be adjustable while securing it to the seat post. Therefore, the seat angle and height are variable within limits. In inexpensive bikes, a simple nut-and-lock washer arrangement permit the biker to alter the seat position from horizontal to a slightly upward or downward angle. Note Fig. 3-1.

Various factors determine seat adjustment. Will the bike be used for professional road racing? Or will it be used for pleasurable touring, where comfort is more important than speed? What is the age and

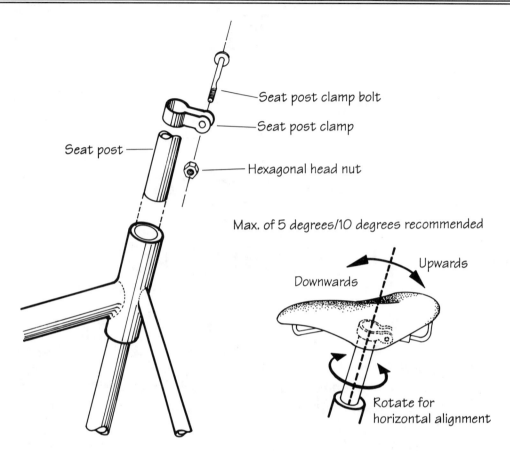

Fig. 3-1. Standard-shaped triangular saddle with angle adjustment.

physical condition of the operator? What are the biker's likes and dislikes? Figure 3-2 shows different seat angle and height adjustment possibilities.

Those persons involved in road racing and ATB (all-terrain bike) events or tournaments are very concerned with seat angle and height, so that maximum force can be transferred from their legs to the pedal mechanism. Therefore, their bike saddles are set relatively high and are positioned to suit their individual anatomies. Observe the top of the seat tube. At the opening where the saddle post penetrates this tube is a saddle clamp, bolt, and nut-and-lock washer assembly. When this nut is "backed off" or loosened, the seat and post assembly can be raised or lowered within the limits of the post.

Saddles & steering mechanisms 31

Fig. 3-2. Correct saddle height from pedals.

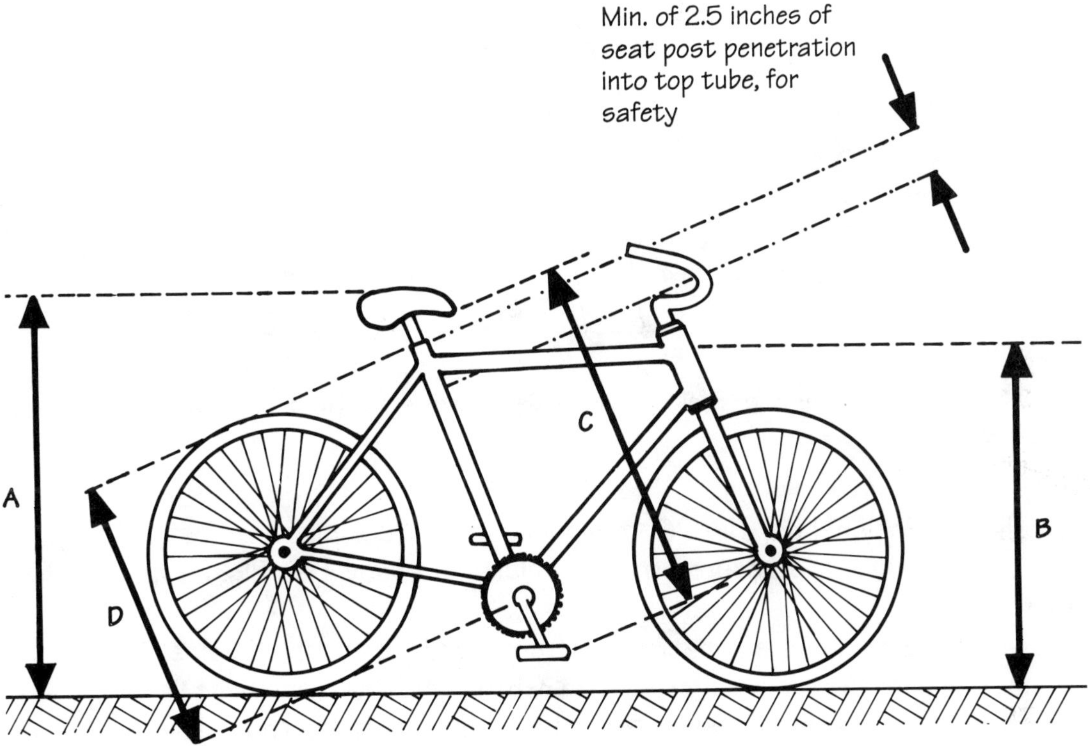

Min. of 2.5 inches of seat post penetration into top tube, for safety

A— top of saddle to ground.
B— top tube to ground—1.5-inch clearance between pelvic bone and top tube. Shoes removed.
C— 1.09 leg length (bottom of foot to inseam (crotch) measurement.
D— varies per individual anatomy and age, hence size of biker.
Frame size—17-inch minimum
26-inch maximum
Inseam measurement—26 inches to 30 inches minimum
35 inches to 38 inches maximum

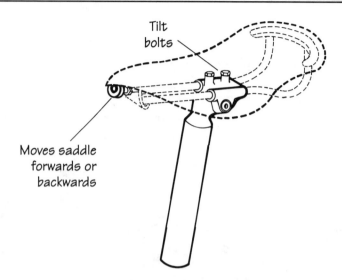

Fig. 3-3. Saddle attachment to post. Details of Campagnola's aluminum alloy, micro-adjusting model.

For the professional bicyclist, Campagnolo produces a micro-adjusting saddle post manufactured from an aluminum alloy. See Fig. 3-3, above.

Seats or saddles are constructed from a one-piece, molded plastic material or are formed from leather, in which a top piece is stitched to an edge piece. Some seats have small air holes that allow air circulation through them so that moisture can evaporate.

Saddle shapes vary. Road racers have the slimmest configuration for minimum weight, not comfort. They are, of course, triangular, and like most bikes, have two individual coil springs in their attachment design. Bicycle Corporation of America (BCA) manufactures three basic design types: anatomic, hard plastic, and vinyl plastic. BCA's racing models have the slimmest triangular configuration. Refer to Fig. 3-4.

Fig. 3-4. BCA saddle designs.

Joe Breeze Cycles, 18 Meadow Way, Fairfax, California, manufactures an excellent saddle for those involved in mountain biking. Sold under the trade name of Hite Rite, this patented seat locator permits the biker, while riding, to adjust his seat quickly and easily. On descents or rough trails, experienced riders depress their seats for a lower and safer center of gravity (C. of G.) and for better control. When the rider returns to level terrain or climbing attitude, the seat springs back to its original preset position. Five different models are available, each recommended for different adjustment needs. Figure 3-5 displays the Hite Rite model with a 2½-inch movement spread. Note that the seat is constructed of molded plastic.

Fig. 3-5. Breeze Hot Rite mountain bike saddle.

Three-wheelers are often designed for seniors. At Lakeside Village Mobile Home Park, Hobe Sound, Florida, where I spend my winters, I frequently see 85- to 90-year-olds driving three-wheelers. In these models, seats are designed primarily for safety and comfort. Trailmate of Sarasota, Florida, produces numerous models, all with these factors in mind. The EZ-Roll model (Fig. 3-6) has a high-back bucket seat; it is hard to fall out of this one! The Desoto Classic has a western-style saddle. Good bikes and good marketing demand unique and clever names (Fig. 3-7).

Fig. 3-6. Trailmate's EZ-Roll tricycle with high-back bucket seat.

Fig. 3-7. Trailmate's Desoto Classic tricycle with western saddle-style seat.

Trailmate also manufactures trikes having molded-plastic, cushioned rectangular seats and oversized triangular ones. For shock absorption, some models have two coil springs; others, a single shock-absorbing cylinder at the seat's front edge and a steel-tubing U-shaped member at the seat's rear. See Fig. 3-8.

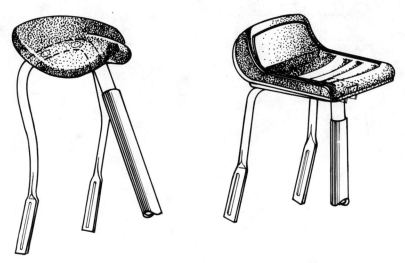

Fig. 3-8. Trailmate's trike seats, showing design features.

Manufacturers of recumbent bikes, those on which the rider comfortably sits with legs at approximately a 90-degree angle to hips, utilize seats made of a strong, nylon-mesh fabric secured to a steel tubing frame. Kann Manufacturing Corporation of Guttenberg, Iowa, produces an aluminum-alloy recumbent bike known as the Linear. Figure 3-9 shows one model. An adaptation of the recumbent is the Funcycle, produced by Trailmate. Body movement is translated into "riding the rail" on this low-slung, fun machine (Fig. 3-10). For obvious safety reasons, this bike is not recommended for highways.

Cycle-Shock: Shock-Absorbing Seat Post

A firm located near Ottawa, Canada, is now manufacturing a shock-absorbing seat post that can be adapted to most bicycle frames.

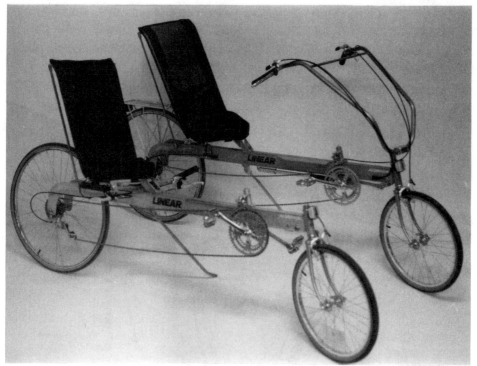

Fig. 3-9. Kann recumbent bike Linear model, with chair-like seat.

Fig. 3-10. Trailmate's Fun Cycle recumbent bike.

This unique item, patented in both the U.S. and Canada, has many important bicycling features, in addition to shock-absorption:

- Reduction of jarring to the human spine
- Reduction of stress to the frame and rear wheel
- Increase of riding comfort and safety
- Improvement of bike control
- Elimination of rear wheel hop
- Fabrication from space-age materials
    - ~ outside tube: 6063-T8 duralumin
    - ~ inside plunger and rod: 6061-T8 duralumin

Because the seat post's patented design will decrease the incidence of spine and related injuries, Cycle-Shock has been endorsed by the Chiropractors Association of Eastern Ontario.

The Cycle-Shock is moderately priced, $70 in Canada and less in the U.S. Figure 3-11 illustrates this unique shock-absorbing seat post.

Fig. 3-11. Cycle-Shock, a shock-absorbing seat post.
Northwings, Inc. Group

Steering Mechanisms

The handlebars for a single-speed, middle-weight bicycle are parallel to the normally horizontal top tube. An exception is the top tube of the Trailmate Easy Rider Classic, angled downward, as is the hand portion of the bar. See Fig. 3-12.

38 *Psyched on Bikes*

Fig. 3-12. Standard one-speed midweight bike, with conventional handlebar — Trailmate's Easy Rider classic.

Multispeed, lightweight, derailleur-type road racers are equipped with curved-down, drop-type bars, which allow the operator to extend the maximum driving force to the pedals—not particularly comfortable, but necessary to maintain maximum forward velocity.

Motocross or BMX bicycles have semi-highrise handlebars. These bikes have 20-inch diameter wheels, in lieu of the normal 26 inches, and are useful for performing quick turns and other maneuvers. Figures 3-13 and 3-14 provide examples of these specialized bikes.

The freestyle bike has the ultimate upwards dimension in its handlebar design, even greater than the highrise model. This freestyle bike is for advanced professional tricksters. Not my cup of tea! Its wheels also have large mag-style spokes—usually five equally spaced ones—similar to those in the BMX machine.

Instead of the horizontal U-shaped bars found on single-speed machines, mountain bikes have a single, robust steering tube. The tube forms a 90-degree angle with the post. Note this design in Fig. 3-15. The mountain bike provides the rider with solid road grip and a quick response to turning—invaluable qualities for biking on extremely rough terrain.

Saddles & steering mechanisms 39

Fig. 3-13. Multispeed derailleur, with drop handlebar on Wheaties/Schwinn road racer.

Fig. 3-14. Highrise and BMX handlebar design.

Fig. 3-15. Mountain bike: KHS's Montana Comp, with single, robust steering tube.

Trikes or tri-wheelers are equipped with handlebars similar to the single-speed, middle-weight bicycle's shown in Fig. 3-12, or the handlebars may be higher, as in Fig. 3-16.

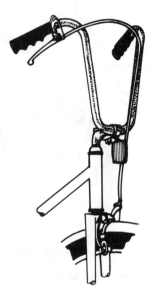

Fig. 3-16. Trike handlebar design.

Tandem bikes, such as the Trailmate in Fig. 3-17 and the Burley in Fig. 3-18, use different steering mechanisms for the lead and second riders. Trailmate's conventional low-rise, U-shaped handlebars are ideal for pleasure biking. The Burley bike, the Duet, is both a touring and sport-riding machine. It has won national tandem events in the U.S. and has been a top contender in European races. The Duet sports a lightweight frame and has down-curved, drop handlebars for the lead rider.

Fig. 3-17. Trailmate's Easy Ride tandem, with low-rise, U-shaped handlebar.

Handlebars for the second rider, or stoker, also vary. Trailmate's Classic and Easy Ride tandem bicycles use a semi-highrise design to provide more comfort than speed. Again, refer to Fig. 3-17. The Burley designs make use of an upwardly curved stoker handlebar secured to a slightly upward-angled tube, or a simple hand grip secured to the front seat post. Figures 3-18 and 3-19 show these design features. The Trailmate and Burley bikes are first-rate machines. Because of their different function and design features, they cannot be honestly compared, and their purchase prices vary considerably.

Fig. 3-18. Burley's Duet tandem, with drop-type lead rider handlebar for sport and touring.

Fig. 3-19. Burley's handlebar design on Samba tandem, a town/country model.

The handlebar design for a folding or collapsible bike is relatively flat. When the bike is collapsed, it must be compact enough to be transported in a car trunk or stored in a large suitcase. The Montague Biframe is an excellent, full-size, high-performance, all-terrain bike that folds into half its operational size within seconds, without tools. It might be costly, but it's one helluva bike! No holds barred! Feast your eyes on Fig. 3-20. For more information, write to Montague Biframe, P.O. Box 118, Cambridge, MA 02238.

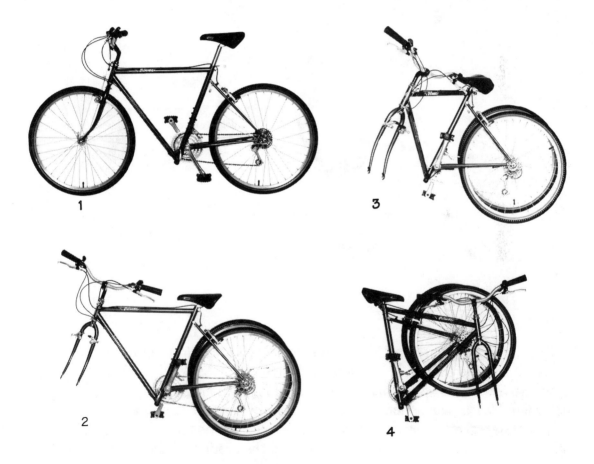

Fig. 3-20. Montague Biframe folding bike.

The backs and upper legs of recumbent bike riders form an approximate 90-degree angle. The steering mechanism must be adaptable to their arm length in this position. As shown in Fig. 3-21, the handlebars have a high rise and a welded crossbar between the U-shape top and the extended hand portions.

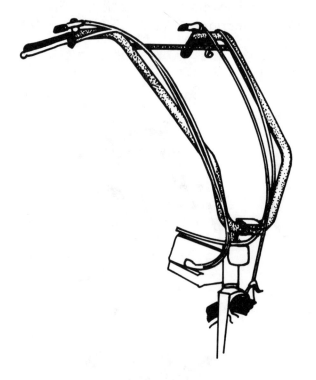

Fig. 3-21. Recumbent bike handlebar design.

The American National Standards Institute (ANSI) in the United States and the Canadian Standards Association (CSA) in Canada conduct stringent, scientific tests on many consumer products, including bicycle components. Their stamps of approval assure the public of safety and durability. If a tested product is not approved for public use, the manufacturer is notified, and the product design goes back to the drafting board. As a consequence, manufacturing firms are under intense financial pressure to produce high-quality products with specifically designed safety features.

Summary

Saddles

- Saddle and seat are interchangeable terminology.
- Most seats are triangular-shaped because of human anatomy; road racing machines have the narrowest wedge shape.
- All saddles have at least two coil springs. Saddle designs vary according to bike function. Trikes offer the softest ride; road racers, the least comfortable.
- Trikes manufactured for seniors have an oversized triangular or rectangular-shaped seat. Sometimes, a bucket design with a high back is used. Again, comfort is stressed.
- Joe Breeze Cycles make a special auto-adjusting Hite Rite saddle for ATBs.
- Recumbent bikes are manufactured by Kann of Iowa and Trailmate of Florida.
- A Canadian firm manufactures an inexpensive shock-absorbing post, adaptable to most bikes, under the trade name, Cycle-Shock.

Steering Mechanisms

- Regardless of bike functions or handlebar shapes, steering mechanisms are manufactured from high-quality, alloy steel tubing, and might be chrome-plated.

- Shapes fall into the following categories:
    - *Single-speed mid-weight*
      U-shaped, but formed in one plane only.
    - *Multispeed derailleur*
      Curved down or drop.
    - *Motocross or BMX: Highrise*
      Trick-performing bikes with semi-highrise bars.
    - *Freestyle*
      Professional trick-performing bike.
      Ultimate in highrise design.
    - *Mountain*
      Rugged, single-tube handlebar.

- *Trikes*
  Similar to one-speed or semi-highrise.
  Configuration provides maximum arm and shoulder comfort.
- *Tandem*
  Lead rider—low-rise, U-shaped, or drop type; depends on bike function.
  Stoker or second rider—semi-highrise or upward curved model.
  Simple hand grip model.
  Depends on bike function.
- *Folding*
  Flat shape in one plane only.
  Takes up minimal space when folded.
- *Recumbent*
  Very highrise, with long, extending 90-degree hand portions.

CHAPTER FOUR

Transmission systems

During the past two centuries, resourceful and inventive people have attempted to use their leg-muscle power to transport them on the ground, on the water, and through the air. The Penny Farthing bicycles that appeared in England during the 19th century (Fig. 4-1) molded plastic paddleboats of the past few decades, and more recently, a one-man aircraft that uses a modified bicycle transmission are examples of these human endeavours. The ultra-light-weight Gossamer Albatross airplane was able to cross the English channel—non-stop—because of an athletic pilot in top physical condition and his skillful use of the expertly modified transmission system. Accolades are certainly due to the plane's American designers and builders.

Not to be outdone by their U.S. neighbors, some Canucks (Canadians) designed a two-man airplane driven by two propellors and a simularly modified bicycle transmission system. The Canadians, led by Arthur le Cheminant, formed an organization known as the Ottawa Man-Powered Flight Group. Their chief designer, Polish-born Waclaw Czerwinkski, helped design Canada's AVRO CF-100, and the AVRO-105 Arrow. In the 1950s, the Arrow was 15 years ahead of its competitors, but the conservative government at that time killed the project, apparently because of enormous cost overruns. Figures 4-2 and 4-3 portray these man-powered airplanes.

Fig. 4-1. Penny farthing bicycle.

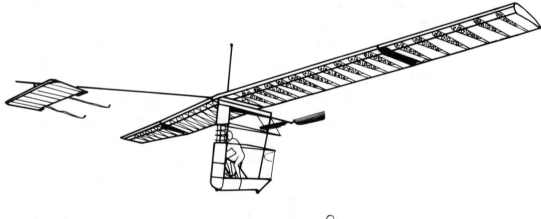

Fig. 4-2. Gossamer Albatross airplane.

Transmission systems   49

Fig. 4-3. Ottawa M-P Flight Group airplane.

Pedal Mechanisms

Pedal Cranks

To harness leg-muscle power to propel a bicycle, you need a suitably engineered combination of wheels, sprockets, chain, pedals, and pedal cranks. Each component is of equal importance. The lack of any hinders proper bike operation. Let us now consider the pedal crank component.

1. *Pedal crank integral with crank shaft*—See Fig. 4-4. This component is found in the crank hub. It is machined-forged and fully removable. Its advantages include design simplicity, low cost, and ease of maintenance. The cranks cannot work loose; they are a forged part of the shaft, which rotates in ball bearings held in the crank hub. Inexpensive coaster-brake bikes, designed for beginners, frequently use this component.

2. *Pedal crank, individual and removable*—This design allows individual cranks to be attached to the crankshaft. Each crank is held to the opposing side of the freely rotating crankshaft by an alloy steel taper pin, hexagon nut, plain and lock washers (Fig. 4-5).

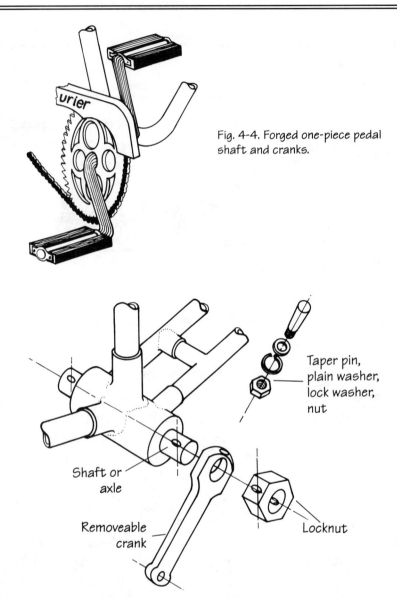

Fig. 4-4. Forged one-piece pedal shaft and cranks.

Fig. 4-5. Simplified view of a forged shaft and removable cranks, with taper pin details for the same.

Occasionally, the hexagon nut will work loose and fall off. Subsequently, the taper pin might also work loose. Fore and aft "play" in the crank in the rotational direction of the main sprocket will become evident. To avoid this problem, you should occasionally check to see if the fastening nut is secure. Because the pin is tapered, it can only pass through the crank and shaft in one direction. Do not attempt to drive it in from the opposite or wrong side with a hammer. Not only is the pin tapered but so is its mating hole.

3. *Shimano Dura-Ace crank and wheel (drive-sprocket) mechanism*—This well-engineered design has a crank that is integral with the drive-sprocket spider. Sprocket and spider are securely riveted together into an integral unit. A simple and effective high strength-to-weight ratio is the result. The opposing pedal crank is fastened to the shaft mechanism located within a polycarbonate-resin drive sprocket hub (Fig. 4-6). This particular design is normally intended for professional—and would be professional—racers. Finely machined of high-quality alloy steel, Dura-Ace or similar mechanisms in a bike involve a substantial financial outlay. Depending on your goals, the investment might be worth it. Only you can decide.

Fig. 4-6. Shimano's Dura-Ace crank and wheel, and other transmission components.

Pedals

Once again the cost factor enters the scene. Bicycles are not unlike cars or motorcycles: You get what you pay for.

Inexpensive ($100 to $150) bicycles have adequate rubber pedal pads, "so-so" metal stampings, and bearings of a sort. With youngsters, the life of this equipment is limited—depending on the aggressiveness of the individual young person. However, replacement of a complete bicycle-pad assembly will not break your bank account.

Some not overly expensive assemblies have better-quality bearings in lieu of the run-of-the-mill ones. Instead of rubber pads, some manufacturers use a composite material with a longer life. In general, when you make your initial purchase, look at the gauge or thickness of the metal used for the pedal stampings, the ball bearings and their races, and the quality of the chroming. Don't be penny wise but pound foolish!

For those who are involved in racing or who desire the ultimate in equipment, Shimano and others offer a variety of precision-engineered and manufactured top-quality pedal assemblies, with or without toe clips. The cost of these units is much higher. Figure 4-7 shows a Shimano pedal assembly.

Fig. 4-7. Shimano PD-7401 pedals.

Friction

As we roll along the byways and highways of America, we desire that all moving parts in our bike offer the least resistance to friction. The contact areas between the tires and the pavement should be sufficient to provide good traction and no more. On stopping, we want maximum friction between brake pads and rim (or in a coaster brake, the tightening of the split cones in the hub), adequate to bring our machine to a full and safe halt, as tires and pavement talk to each other.

Ball Bearings

The balls in this antifriction device are confined in a thin steel race, but are free to rotate in their separate cages within the race. Inexpensive and mid-priced machines use ball bearings in front and rear axles, in drive sprocket hubs, and in pedal assemblies. If any balls fall out of a race, you should replace the unit before more serious damage results. If you must take apart the sprocket hub, you should replace the race assembly with an identical part. Replacement is often necessary when the bike has been driven consistently over sandy beaches or roads. Whenever your bike is sand-or mud-coated, never hose it with water—worse still, with water and detergent. Bearings were never designed to take that kind of punishment. Instead, wipe down your bike with a cloth dampened in oil-based paint thinner, such as Varsol. Use a dry cloth to finish the cleaning procedure.

Roller and Needle Bearings

Small steel alloy rollers can be used instead of balls to reduce friction. This needle-type bearing is simply smaller in diameter than its cousin. Once again, Shimano of Japan uses both types of bearings in its PD-7400 and PD-7401 series of road racer pedals. These bearings are sealed to prevent contamination from road dust and water. When the dollar stakes are high, such as in the Tour de France competition, fractions of a second mean the difference between winning and losing. Consequently, international competitors use the best. Figures 4-8 and 4-9 illustrate this engineering concept.

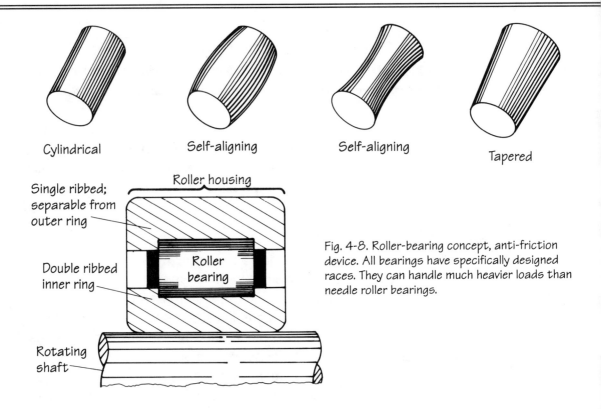

Fig. 4-8. Roller-bearing concept, anti-friction device. All bearings have specifically designed races. They can handle much heavier loads than needle roller bearings.

Bearing Pictorials

1. Much smaller in diameter than roller bearings, hence cannot bear a heavy load.

2. This series of needle bearings retained by an outer ring and separate inner ring. Cut-away drawing not shown.

Cutaway Illustration

One-piece channel-shaped, roller guiding outer race only.

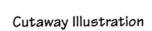

Fig. 4-9. Needle-bearing concept, anti-friction device.

Lubrication & Other Maintenance

Inexpensive and mid-priced bikes may require more lubricating than expensive track-and-road bicycles. But the expensive ones, too, have special lubricating parts with "open/close" indicators on necessary components. Sealed components—properly lubricated at the factory—are also used in multiple free-wheel mechanisms for the same track-and-road machines. Racing crew specialists, using their favorite lubricants, have their own formulas for frequency of lubrication and quantity used. S.A.E. (Society of Automotive Engineers) No. 40 oil from a quality bicycle or automotive shop will do the job adequately. Tubular oil ports with spring-loaded caps; or a spring-loaded ball-bearing sealed in an oil hole; or simply, a small oil port or hole that can be covered quickly with a metal clip that rotates about the axle hub, are all used in the lubrication process. The quality or cost of the bike will dictate which style is used. The name of the game is to circulate the oil in and around the bearings, while preventing water and grit from contaminating them. Lubrication frequency is determined by the number of miles or kilometers a bike is ridden. Weather conditions—rain and high temperatures—amount of use, and terrain conditions all have a bearing on the oiling schedule.

    A light oil such as S.A.E. 40, should be sparingly applied to the drive chain. This task is not difficult if a second person lifts the rear wheel off the ground while you slowly rotate the pedal crank and apply the lubricant. If a second person is unavailable, turn the bike upside-down, resting it on its handlebars and seat. Carry out the same lubricating procedure, as above. The smaller-diameter front wheel hub and the larger-diameter rear wheel hub, containing either the coaster-brake mechanism or internal gear-shifting mechanism, should also be lubricated as necessary.

    The drive-sprocket hub will also need oiling attention periodically. However, as stated before, oil sparingly. Smaller quantities of oil applied frequently are better than larger quantities applied infrequently. Why? When you drench bike parts with oil, you get back dirty socks and soiled trousers. You also waste oil. On older bikes, a leather strap was frequently fastened loosely around axle hubs to cut down an oil spillovers. The strap also acted as a buffer and kept the chrome shining.

In conclusion, over-oiling is not necessary, but too little or no lubrication will hasten your two-wheeled investment's ride to the junk heap. Use common sense, but also use some oil, albeit sparingly.

Greasing

In places where oil cannot be applied—and grease will remain in contact—a special water-resistant grease may be used. A very light application to the chain drive might be necessary.

Cleaning

As I stated earlier, do not hose down your bike with water or water and detergents. Should these liquids enter the axle or sprocket housings, bearings will be prone to oxidation and, over a long period, perhaps minute pitting. Pour Varsol or similar petroleum-based paint thinner on a cloth and wipe off dust, oil, and other road contaminants. Clean and wipe dry with another soft cloth. Water is a no-no—for cars, ok; for bikes, no way!

Transmissions

One-speed Transmission, with Coaster Brake

Prior to the design and manufacture of internal gear-shifting mechanisms by Sturmey-Archer of England and Shimano of Japan, and the external derailleur gear-shifting mechanisms by French, Italian, and Japanese companies, only the single-speed coaster-brake mechanism was available. This device is still adequate for preteenagers, but teenagers will hanker for a more sophisticated bike. Still, I recommend to parents and grandparents paying the bills to forgo their youngsters' whims. Until these children have mastered rudimentary safety skills in handling their bikes, a single-speed is more than adequate. Figure 5-1 in the following chapter shows a typical single-speed or coaster-brake component.

The simple maintenance problems inherent in this bike type are minimal. Many parents and some preteenagers should have the aptitude to carry out effective repairs.

The pedal crank arms might work loose due to excessive vibration caused by rough handling such as jumping curbs or dropping bikes onto hard pavements, thus striking the pedals. A suitable wrench is sometimes provided with a new bike and might help solve the aforementioned problem—when used properly. Note that the pedal crank arms, to which the pedals are attached, always rotate in a counterclockwise direction when the bicycle is in operation. The threaded shaft for the left-hand pedal has a left-hand thread, while the shaft for the right-hand pedal has a right-hand thread. If this were not so, the left-hand pedal would continually work loose. Figure 4-10 illustrates the pedal structure. Figure 4-7 shows the excellent design of the Shimano PD-7401 pedals. Note that the machined collars adjacent to the threaded shafts are stamped L (left) and R (right) to alert the repairer to the threaded shaft's direction.

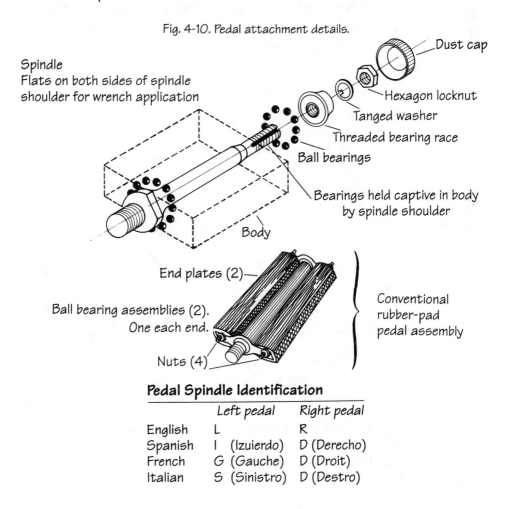

Fig. 4-10. Pedal attachment details.

Pedal Spindle Identification

	Left pedal	Right pedal
English	L	R
Spanish	I (Izuierdo)	D (Derecho)
French	G (Gauche)	D (Droit)
Italian	S (Sinistro)	D (Destro)

Three-Speed Internal Gear-Shifting Mechanism

Both Sturmey-Archer and Shimano manufacture these mechanisms. Shimano has gone an additional step by manufacturing a three-speed hub that includes a coaster brake. The complex design of these devices precludes the majority of owners from repairing them if internal damage or malfunction occurs. However, you might make cable adjustments—either tighten or loosen cable attachment links emanating from the three-speed hub, which sets the different speed positions accurately. Otherwise, take your bike to an authorized dealer or service center for repairs. These mechanisms are well designed and fabricated, so they seldom fail. But over time, the shaft cable between the hand-operated lever and the shift mechanism in the drive hub can become slack. Simple connecting devices permit the owner to correct this minor problem. Figure 4-11 describes this repair operation.

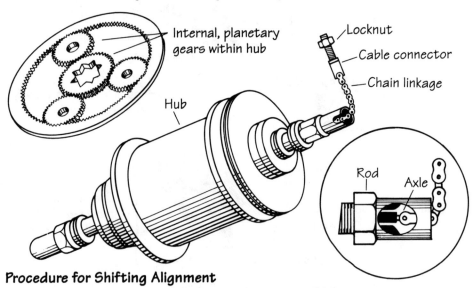

Fig. 4-11. Internal gear-shifting mechanism.

Procedure for Shifting Alignment

1. Move gear control lever on handlebars to second gear position.
2. Loosen locknut.
3. Turn cable connector until shoulder of gear indicator rod is perfectly aligned with end of axle.
4. Tighten locknut. Gears are now in correct alignment.
5. Monthly maintenance—2/3 drops of S.A.E. 20 oil.

Just remember—some minor servicing and repair jobs are within your reach, but know when to say no to jobs you cannot handle.

Multispeed Derailleurs

Derailleur is a French technical term meaning "to derail." In essence, derailing is what the operator of a 10-, 15-, or 18-speed bicycle does. The chain wheel or drive sprocket has two or three sets of gear teeth. These teeth mesh with one gear in a cluster of from five to seven parallel, but much smaller gears or sprockets. By mechanically slipping the drive chain from any one of the two or three drive wheels to any one of the five to seven smaller driven sprockets, you can obtain a very low or very high gear ratio, which allows you to climb hills—high gradients at low speed—or to ride on flat surfaces or downhill stretches—low gradients at high speed.

The slipping or derailing of the drive chain is neatly performed by hand-operated levers. See Fig. 4-12. Once again, Shimano of Osaka,

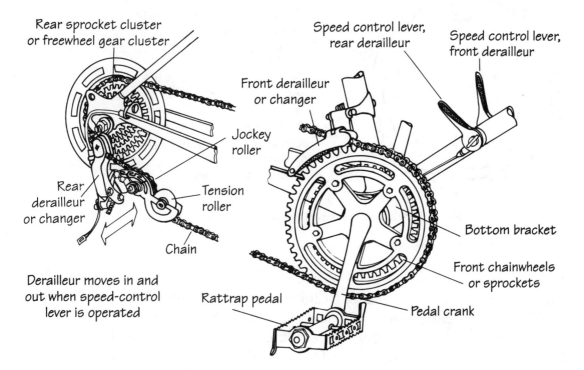

Fig. 4-12. Standard derailleur mechanism.

Japan, has become a world leader with its "Integrated-8 Shimano Index System" of precision-shifting. With this system, you can precisely change or shift gears through hand-operated steps or indexes. You can shift to take a sharp turn as easily as you can ride a straight path. Shimano has used computer-simulated techniques to further improve its chain-wheel (drive-sprocket) design. No longer a perfect circle, Shimano's chain wheel is now egg-shaped, or, as the company describes it, a "point-symmetrical egg curve." (Fig. 4-13). This "Biospace" chain wheel looks simple, but it is not. It allows for better pedaling cadence and efficiency, and a more natural feel. Once again, this top product has a high price tag. For further details, see Figs. 4-13 and 4-14.

Biospace Highlights

Point symmetrical egg curve shape—established by computer analysis and design.
Produces:
1. Lighter and more natural pedaling cadence.
2. Saves energy.
3. You ride farther and more comfortably with less fatigue and knee strain.
4. Hill climbing is easier.

Fig. 4-13. Biospace chainwheel. Shimano American Corp.

Shift Levers

Suntour, U.S.A., Inc., of Novato, California, carries a wide range of shifting components, many of them, no doubt, manufactured in Japan, under the Suntour label. Figure 4-15 shows the Suntour Model SL-SB00 Superbe-Pro shift lever. Figure 4-16 is a detailed photograph of the components for the Superbe-Pro transmission system.

Transmission systems

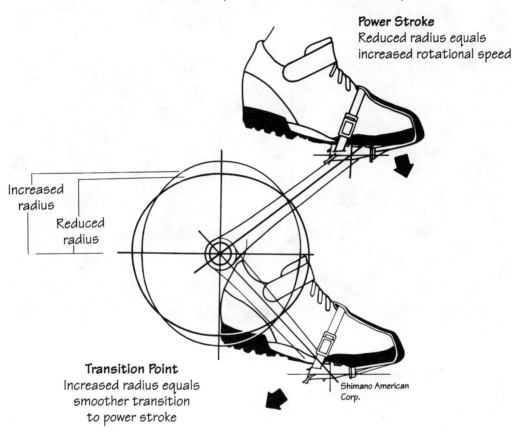

Fig. 4-14. Biospace chainwheel engineering concept.

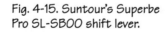

Fig. 4-15. Suntour's Superbe Pro SL-SB00 shift lever.

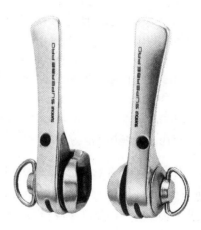

Fig. 4-16. Components of Suntour's Superbe Pro transmission system.

Chain Tension Adjustment

Drive chains seldom jump their sprockets. Inexpensive children's bikes with a coaster-brake design are more prone to do so because of rough treatment. Older bikes had a simple machine screw-operated shackle that slipped over the rear entry slot of the fork, where the rear wheel axle was contained. By adjusting this screw, you could pull the rear wheel back and tighten the chain. The reverse movement would permit chain slack to increase. Single- or multispeed bicycles manufactured within the past few years do not have this chain-tightening feature. With newer bikes, you must loosen the two rear-wheel axle nuts and pull the wheel back until the desired tension is reached. This procedure works, but is technically crude (Fig. 4-17).

Chain Removal and Reinstallation

Apply some pressure to the chain at the top of the chain wheel (drive sprocket) as you turn the pedal crank forward to rotate the crank. If

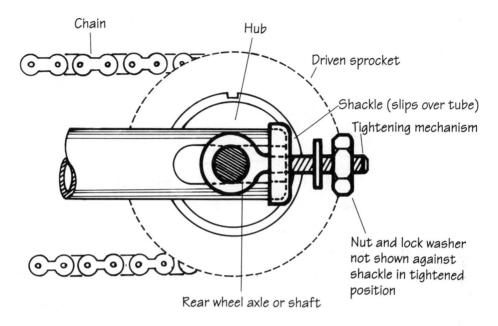

Fig. 4-17. Method of tightening drive chain.

the chain is very loose, it will jump the sprocket without much trouble.

To reinstall a jumped chain, be sure that the chain is fully engaged on all the contact teeth of the rear or driven sprocket. Then, reposition the chain over the top, or uppermost, teeth of the chain wheel and turn the pedal cranks counterclockwise. Hands-on experience will increase your odds of doing this repair quickly. Clean your hands with waterless skin cleanser after this operation. Varsol is an effective cleaning agent.

Chain Link Removal

In many bikes, the chain is not continuous but is positioned by a link, clip, and integral pin arrangement. These arrangements are removable, as can be seen in Fig. 4-18. Too much chain stretching, followed by a link breakage, requires immediate repair. It is not a difficult task, and the tools required are minimal.

However, ATBs, MTBs, hybrids, road bikes, and other quality machines have a riveted, continuous chain that requires a special chain-maintenance tool. See 6-37 in chapter 6.

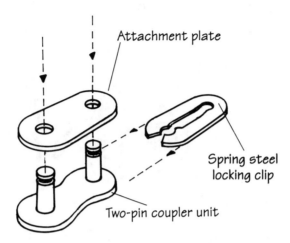

Fig. 4-18. Details of chain—removable link, clip, and integral pin.

Lubrication

1. Use a quality light oil, S.A.E 40, or one of a particular viscosity recommended by the manufacturer.
2. Oil sparingly and occasionally, as mentioned earlier in this chapter.

Summary

Purchase Tips

- A single-speed transmission bike is adequate for a pre-teenager.
- A multispeed derailleur bike is a more suitable choice for the college student or young professional because of its wide variety of shifting mechanisms.
- An older person might be satisfied with the three-speed transmission (internal shifting mechanism) found in a three-wheeler, since high speed is not an important factor. Such bikes—with carriers for shopping purposes—are quite common in retirement communities.

Repairs

- Single-speed mechanisms are not normally prone to misadjustment or breakage. Replacement of a complete pedal-pad assembly is easily carried out. Loose pedal cranks can be simply tightened or

replaced, if necessary. A jumped chain can be fed back onto the drive and driven sprockets. A loose chain can be tightened without serious effort.
- A three-speed mechanism that does not shift gears at the correct lever position can be realigned without much trouble.
- Replacing a frayed or severed shift cable is easy.
- Do not attempt to tinker with the inner workings of a three-speed internal shift mechanism. This repair is a job for a qualified dealer or service center.
- Misaligned derailleur chains can be retracked correctly without great skill. Refer to the handbook accompanying your particular bike for specific details.

CHAPTER FIVE

Braking mechanisms

Every bicycle needs a safe and simple braking system. Traffic enforcers no longer look the other way at bicyclists who disobey traffic regulations. Bicycles—even though operated by human muscle—are still moving vehicles, and law-enforcement officers are now checking them more closely for safe mechanical condition. Be aware! Be alert! Be smart! Not only your life, but the lives of bystanders, pedestrians, and other vehicle operators might be in jeopardy because you have neglected to maintain safe and dependable brakes on your bike.

A Caliper Brake Hybrid

A single-speed bike generally uses a rear-hub coaster brake. However, Shimano does manufacture a three-speed, internal gear-shift bicycle with a coaster-brake mechanism. Normally, three-speed machines come with caliper brakes. Obviously, there must be a consumer demand for this hybrid.

With normal pedaling, no braking occurs if the bicyclist rotates his pedals in a counterclockwise direction. The bicyclist must exert quick adequate force to the right pedal and crank in a clockwise direction for the rear wheel to stop turning. Some daredevils driving bikes with this type of brake often skid intentionally or perform other unsafe maneuvers. Rim-type caliper brakes do not respond well to braking or provide the force required for skidding. As a result, many youngsters lose fair amounts of skin from legs, knees, and hands because of skidding and performing wheelies. The damage to their bikes can also be substantial. But "youth must be youth", so the adage goes!

Coaster Brakes

Figure 5-1 shows a line drawing of a typical coaster brake. With normal pedaling, both the lever and brake cones do not put pressure on the brake sleeve. To generate forward motion, power is transmitted from the drive chain to the drive cone to the drive sprocket and finally, to the rear wheel. When a bicyclist coasts and doesn't pedal, neither braking nor motive action occurs.

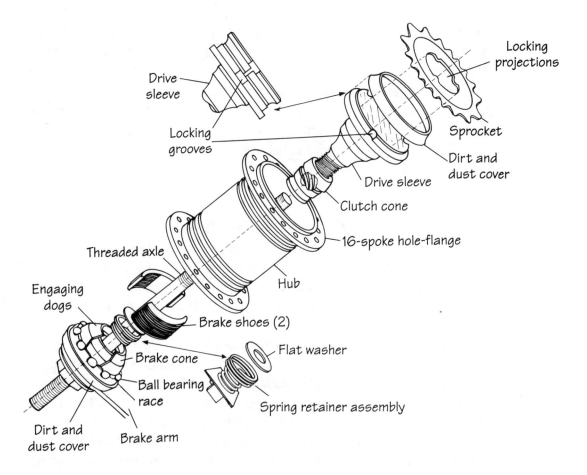

Fig. 5-1. Coaster-brake mechanism.

Cones, sleeves, and the roller-guide ring can develop cracks, finally split, or become deeply scarred from road contaminants such as fine sand. Such damaged parts must be replaced as soon as possible. A youth or adult with reasonable mechanical skills should be able to disassemble and reassemble coaster-brake parts.

The drawings in this book should help. Still, if a manufacturer's guide booklet is part of the bike package, so much the better. Make sure all parts have been cleaned of road grime. Remove damaged parts and replace them with identical new parts (check the manufacturer's part numbers). Make sure that the replacement parts are in the same position as the parts being removed.

Notice the hexagon nuts at the end of the threaded axle shaft on which the assembly is made. Provide a little space between the ends of the assembly and these nuts; otherwise, frictional jamming will take place. These hexagon nuts do not keep the driven sprocket and hub assembly in the fork slots. Two additional nuts and lock-and-plain washers perform this function. Also, make sure that the brake lever is properly engaged to the nonsprocket side of the hub assembly and is securely attached to the frame with the U-clamp, machine screw, and the nut and lock washers.

Finally, make certain that the hub assembly is properly lubricated. Then reattach the drive chain, and test the mechanism for motive free-wheeling and braking actions.

Caliper Brakes

Caliper brakes are externally mounted to the bicycle's frame and operate against the rim of both the front and rear wheels.

When the operator squeezes a hand-operated lever, the caliper encircles the wheel rim. These brake levers are fastened to the handlebars, along with strong braided steel cables. The left-side handlebar lever operates the front-wheel caliper brake. The right lever controls the rear-wheel caliper. These cables are secured to an arm that operates the caliper. When the caliper closes it brings a rubber braking pad into contact with the wheel rim, just below the tire. Refer to Fig. 5-2 for details of a typical caliper brake of the side-pull type. Another side-pull design—the Shimano Super SLR can be seen in Fig. 5-8.

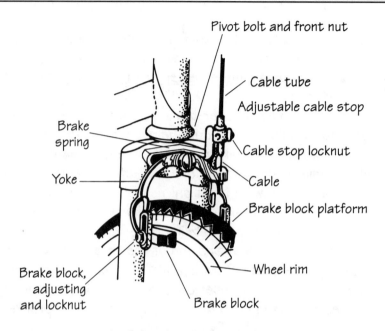

Fig. 5-2. Side-pull caliper brakes.

Caliper brakes are of three designs: side-pull (Fig. 5-2), center-pull (Fig. 5-3), and cantilever (Fig. 5-4). A center-pull caliper probably does not achieve important advantages over the side-pull type and is probably more expensive to produce.

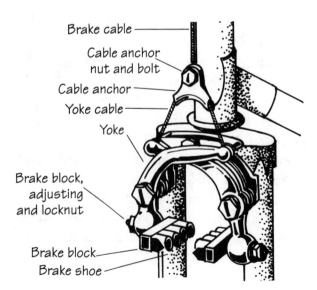

Fig. 5-3. Center-pull caliper brakes.

Braking mechanisms 71

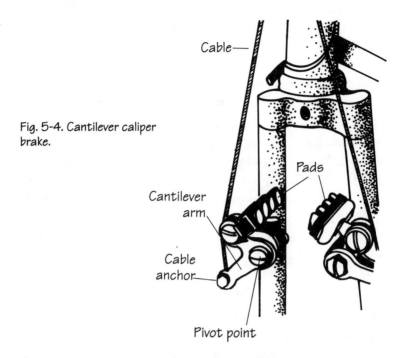

Fig. 5-4. Cantilever caliper brake.

The apparent advantage of the cantilever caliper brake is the greater leverage obtained because of the long arms pivoted to weldments attached to the fork tubes. Consequently, the biker can achieve more stopping power with less braking effort than he or she could obtain from a center- or side-pull caliper brake. Note the cantilever arms in Fig. 5-4.

Since braking action depends on the surface area of the pads making frictional contact with the steel rim, rainy or icy conditions do not auger well for caliper brakes. Keep this in mind and act accordingly. When you ride a bike with these brakes in northern winter conditions, wear helmets and other protective equipment. Certain American states and Canadian provinces have regulations regarding protective equipment. Even if they do not, use common sense, wear such equipment when necessary, and stay alive.

The chances of a skid with a bike having caliper brakes are probably slim. Both brake levers should be operated in unison. Greater braking force should be applied to the rear wheel, rather than the front, when stopping.

A caliper brake system is relatively easy to maintain. First make certain that a ⅛-inch (or 3 to 4 mm) gap exists between the brake pads and the wheels while the wheels freely rotate. You should feel no drag against the rim. Wheels that do not track properly or wobble because of spoke loss will not provide a proper braking surface. The same is true of incorrectly tightened spokes, which cause rim warping. Both conditions can be caused by the rider's jumping concrete curbs and other rough usage. Badly worn pads will not provide good braking action either. If the pads are worn, replace them. On occasion, check the cables for fraying and possible breakage. Replace these when required. Proper kits are available from your supplier.

You can also make adjustments to the lever to reduce or increase hand pressure.

Brake Levers

A select few manufacturers produce well-engineered, high-quality brake components. Shimano and Suntour, U.S.A., Inc., are two of the best. Both are Japanese companies with bases in California. Suntour's Superbe-Pro Model BL-SB00-N brake lever is shown in Fig. 5-5. Both the photograph and an exploded engineer's drawing illustrate this quality unit.

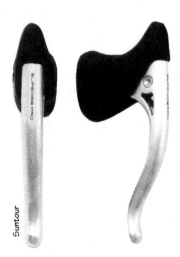

Fig. 5-5. Suntour's Model BL-SB00-N brake lever.

Additionally, a front-wheel brake-lever system, the Suntour X-press, is shown in Fig. 5-6. This excellent unit is mounted on a mountain bike. Notice the straight handlebar and the knobby tire.

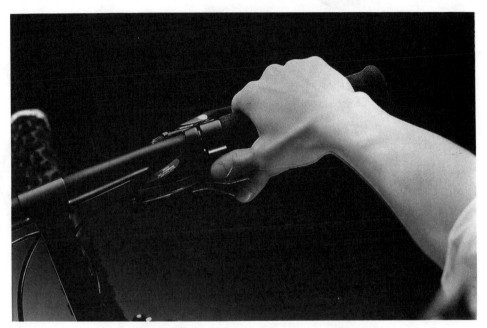

Fig. 5-6. Suntour's X-press brake-lever system.

Finally, if you lose your bike's braking function—i.e., the chain jumps the sprocket of the rear wheel, or the brake cables snap on your three-speed or derailleur—do not panic! You still have feet for walking—always wear suitable shoes, not bare feet or sandals. Get off your bike and guide away from a potential collision situation, or first, slow your speed by gradually applying foot pressure to the pavement. Then get off. A runaway situation is no fun. You might also feel more confident if you are wearing an approved helmet and other protective clothing.

Figure 5-7 shows two recent-model Schwinn bicycles. The background one has a standard coaster brake; the foreground one, a side-pull caliper brake. Note that the front wheels are not used for braking

Fig. 5-7. Schwinn's Phantom coaster-brake model (in background).

applications. The drawing in Fig. 5-8 depicts a state-of-the-art caliper-braking mechanism, a Shimano Super SLR caliper brake. It provides more braking power with less lever input.

Disk Brakes

At least two models are available with a disk-and-drum configuration. Phil Wood of San Jose, California, manufactures a high-quality disk brake, designed for adult tandems. Obviously, it has good stopping features. Shimano's disk model is not strong enough for an adult's weight—especially true if it is a tandem, but its design is intended for the juvenile market.

Drum Brakes

Shimano also produces the Radiax Fin, again designed exclusively for the children's market. Its braking power is limited, and its weight is

Braking mechanisms   **75**

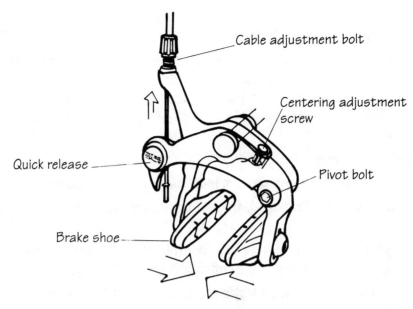

Fig. 5-8. Shimano Super-SLR caliper brake. Shimano American Corporation

heavy—20.8 ounces for the brake alone, and a total of 2.5 pounds with the hub and brake as a unit.

From Sandnes, Norway, the firm of Johnas-Oglaend manufactures a drum model. It is weighty and cumbersome, but needed on a heavy tandem. The positive features of the J.O. drum brake are worth spelling out: an internal expanding brake, excellent in wet weather; a rim that does not heat up, as in a caliper type; and therefore, less brake fadeout.

Summary

Coaster Brakes
- Used primarily in children's bikes and lower-priced adult machines. Foot- and leg-operated.
- Virtually maintenance free. Brake arm must be secured properly to rear wheel and chain stay for good braking conditions. Occasional light oiling of rear wheel hub mechanism is required to maintain smooth operation.
- Unfortunately, this brake mechanism permits brake jamming and rear wheel skidding. Parental guidance can deter this action.

Caliper brakes
- Hand-operated by using handle-mounted lever(s). Skidding is less likely. Braking is a more progressive action.
- Front and rear wheels are normally braked in unison, but with more force applied to the rear wheel lever, if the front wheel has a braking mechanism.
- Repairs are minimal:
  ~ Hand lever adjustment to suit individuals can be made.
  ~ Replacement of stretched, frayed, or severed cables is not difficult.
  ~ Adjustment and centering of caliper might be needed as the rubber pad wears; pad replacement might be necessary.
  ~ Cleanliness, wheel rim condition, and brake contact surface are important to maintain good braking.
- Trueness of each wheel with respect to brake pad clearance is of prime importance.

In either braking system, the operator should be aware that the contact area between tire and pavement is small. An eight-year-old child may be the operator, if not the actual purchaser. That eight-year-old youngster could end up in a serious accident, if unprepared to handle a difficult traffic situation. That eight-year-old could be your son or daughter!

CHAPTER SIX

Accessories

Many bicycle accessories are available in today's market. Some are required by state, provincial, or local traffic laws. Bike riders must make themselves aware of these regulations. For example, by not wearing an approved safety helmet, you might not only be breaking the law, but jeopardizing your life!

Helmets

Buying a helmet is a vital investment. Three-quarters of all bicycle deaths are the result of trauma to the head. For optimal protection, look for a quality brand with a hard outer shell that will not puncture when struck by rocks and other debris. The inside of the helmet must have a minimum of ½ inch of foam to cushion the head from trauma on impact. The foam should be rigid yet crushable. Avoid brands that have soft foam inner lining. Also make sure that the helmet meets ANSI standards and fits the shape of your head comfortably.

Etto Helmets of Palm Springs, California, offers a helmet that surpasses the ANSI Z-904 standard and is also approved by the Swedish National Board for Consumer Policy. This particular design, known as the Etto Classic, has a patented shell-adjustment system that provides a total fit for head measurements ranging from 20.5 to 23.2 inches.

The helmet weighs only 11.3 ounces and is exceptionally well ventilated. It has an outer shell of tough ABS plastic and an inner polysterene liner with five comfort pads that make contact with the rider's head. Five different colors are available for those who wish to coordinate their helmets with their bicycle clothing. Figure 6-1 shows the unique "key method" of adjusting the internal size of the helmet.

Fig. 6-1. Etto's key-adjusted safety helmet.

Specialized Bicycle Components, Inc., of Morgan Hill, California, also manufactures a sleek, aerodynamically shaped helmet that is featherweight, attractive, and, according to the manufacturer, "may reduce your time trials by a few seconds." The manufacturer puts forward no claim to universal adjustment.

Lights

All 50 states recognize a bicycle as a legal vehicle. Canadian provinces do the same. As a consequence, the use of suitable lights and reflectors at night is mandatory.

Six volt d.c. (direct-current) generators were very popular someyears back and are still available. A d.c. rim generator from Union-Fröndenberg-USA, Olney, Illinois, is clamped to either the front- or rear-wheel forks. It is positioned so that a small drive wheel on the generator shaft impinges against the wheel rim, thereby producing electrical current as the wheel rotates. Its disadvantage is that sufficient current to light the head lamp properly is lacking at slow bicycle speeds. Refer to Fig. 6-2.

Accessories 79

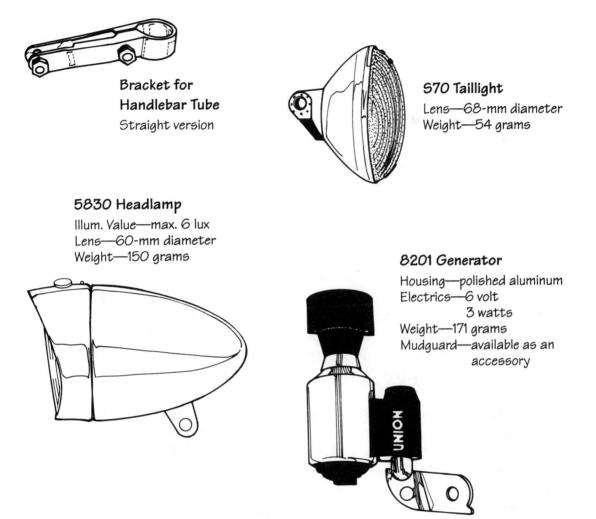

Bracket for Handlebar Tube
Straight version

S70 Taillight
Lens—68-mm diameter
Weight—54 grams

5830 Headlamp
Illum. Value—max. 6 lux
Lens—60-mm diameter
Weight—150 grams

8201 Generator
Housing—polished aluminum
Electrics—6 volt
　　　　　3 watts
Weight—171 grams
Mudguard—available as an
　　　　　accessory

Fig. 6-2. Union-Fröndenberg 6-volt d.c. generator.

Chargeable and nonchargeable battery-powered units are more popular today. Brite Lite Cycling Lights of Soquel, California, produces a variety of excellent light systems, including the Hilighter helmet light (Fig. 6-3). This particular unit can be mounted to either soft- or hard-shell helmets. This feature is particularly useful to free your hands for bike repair at night. Both 1.5- and 4.0-hour ultralight battery packs that produce 4.5 watts of electricity are available.

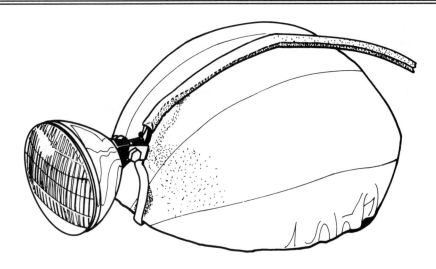

Fig. 6-3. Brite Lite's helmet-mounted light.

Brite Lite manufactures five light systems: Alphalite, Betalite, Standard, Performance, and Super. These range from 2.4 to 10.0 watts of electrical output. Battery pack weight varies from 15 to 36 ounces; and battery time, from 1.3 to 4 hours. Beam patterns up to 100 feet are found in the Standard, Performance, and Super systems.

These other features make the Brite Lite systems top performers: Halogen head lamp; superior multizone beam pattern; sealed, lead-acid, rechargeable battery pack; and high-visibility taillights. The four components—battery pack, charging unit, headlight, and taillight—are shown in Fig. 6-4.

Union-Fröndenberg-USA of Olney, Illinois, also manufactures a wide range of innovative bicycle components in the U.S. and in Germany. Its system, The Light Works, (Fig. 6-5) is similar to the Brite Lite. This system includes sealed, lead-acid, rechargeable battery units.

The Cat Eye organization of Osaka, Japan, is well known for its high-quality, halogen bulb, alkaline dry-cell battery-operated units. Using two C-size nonchargeable batteries, these lightweight units produce six hours of superior halogen-bulb lighting. Model HL-500 is shown in Fig. 6-6. Other variations of the HL-500 are available to suit different mounting conditions. Red taillights and a conventional

Accessories 81

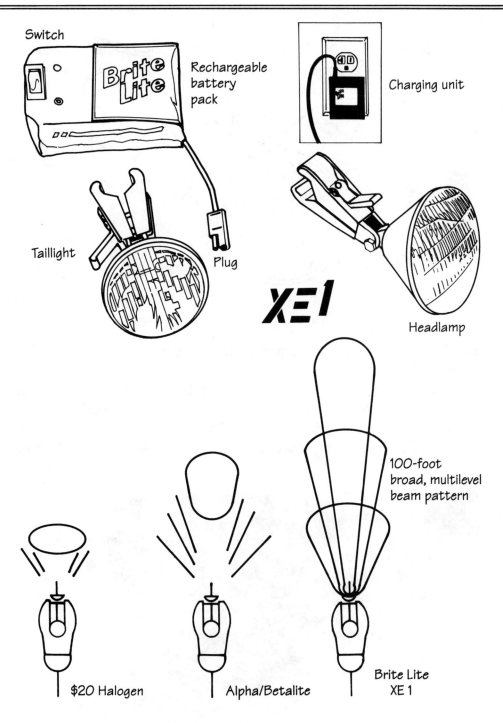

Fig. 6-4. Brite Lite's bicycle light systems.

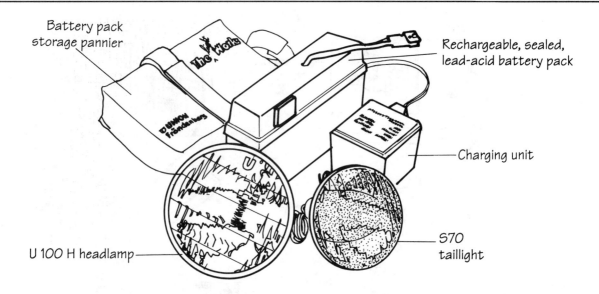

Fig. 6-5. Union-Fröndenberg's The Light Works bicycle light.

quickly dismountable flashlight-type lamp are also part of Cat Eye's product line.

Vistalite, Inc., of Lancaster, Pennsylvania, a relative newcomer in the bicycle accessories field, is making great "light waves" with the introduction of LED (light-emitting diode) bicycle safety lights. Figure 6-7 shows a red taillight attached to a customized bracket, which is

Fig. 6-6. Cat Eye's Model HL-500 halogen bulb, dry-cell bicycle light.

Accessories 83

Fig. 6-7. Vistalite's red safety taillight.

secured to the rear-wheel fork members. The specifics of this new, bright American Star (and other Vistalite models) are as follows:

Vistalite VL-300
- Super luminescent LEDs
- Flashes automatically six times per second.
- Visibility exceeds 2000 feet because of pure red light.
- Rugged, lightweight, totally weatherproof plastic housing (ABS high-impact plastic).
- One-year limited warranty.
- Operates for 300 to 500 hours on two AA batteries, or for 20 continuous days.
- Mounts easily to any existing reflector bracket.
- Lens operates as a wide-angle, prismatic, reflective type when the battery switch is off.

Vistalite VL-300C
- Maintains all the great features of the VL-300, except that it is a clip-on model.

Vistalite VL-310

- Maintains the features of the VL-300 model, except it is exclusively designed as a front bicycle safety light.
- Has five angled LEDs for better side visibility.
- Easily visible up to 1000 feet.
- Slightly larger than regular rear reflector.

These units are not expensive (from $20 to $30); in terms of saving lives, their value lies in the millions. Hats off to Robert Choi and Kwai Kong for inventing, testing, manufacturing, and marketing this American Star.

Bicycle Computers

Cat Eye also produces unique, multifunction bicycle computers. The Micro Model CC-6000, depicted in Fig. 6-8, includes the following features: large, easy-to-read, dual, digital liquid crystal displays (LCD); water resistance; long-lasting 2- to 3-year duration lithium battery; compact, easy-to-install, sensor system; and display unit that can be removed simply and quickly from attachment bracket to prevent theft. This model's functions are as follows: cadence (pedal rotation frequency) = 0 to 199 rpm; current speed = 0 to 65 mph or 105 kph; maximum speed = to 65 mph or 105 kph; average speed = varies; odometer = 0 to 10,000 miles or equivalent kilometers; trip distance = 0 to 1000 miles or equivalent kilometers; elapsed time = 0 to 10 hours; start/stop counter; and mode.

Other differently shaped cyclocomputers are also available from Cat Eye, which has over seven years of specialized experience in this high-technology field. These include the sleek, asymmetrical, aerodynamically-shaped Model CC-CL100, and the CC-8000, designed to be mounted at the center of the steering bar of all-terrain bikes (ATBs). These units are shown in Fig. 6-9. The CC-CL100 is a cordless model; the function information is transmitted by electromagnetic (radio) waves from the sensor to the computer. Model CC-7000, the Vectra (Fig. 6-10) has newly designed "piano-touch," click-feeding buttons for start/stop and mode.

Accessories 85

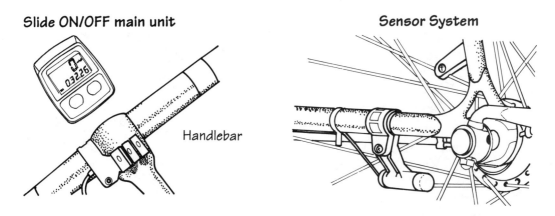

Fig. 6-8. Cat Eye's Micro Model CC-6000 bicycle computer.

Fig. 6-9. The main unit (left) and sensor/transmitter (right) for Cat Eye's Model CG-8000 bicycle computer.

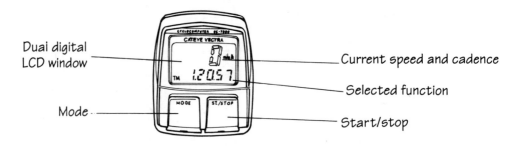

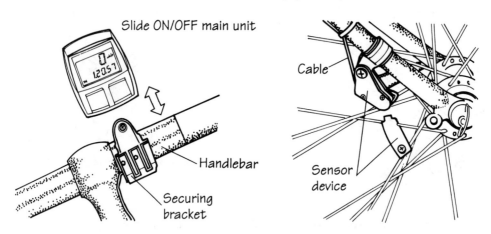

Fig. 6-10. Cat Eye's Vectra Model CC-7000 bicycle computer.

Rearview Mirrors

A rearview mirror is a necessary safety item for all bicyclists, especially when you are operating your machine on congested city streets. Mirrors are either chrome-plated metal or chrome-finished plastic, with a hard coating to eliminate scratches and abrasions.

Tsuyama Manufacturing Company Ltd., Osaka, Japan—the same firm that makes bicycle accessories under the trade name of Cat Eye—also produces mirrors. Tsuyama's racing bike mirror, Model BM-300G (Fig. 6-11) is a 3-inch-diameter chromed plastic unit that attaches simply into the end of a drop-style handlebar. Figure 6-12 displays a more conventional mirror that clamps to the handlebar just forward of the grip. Left- or right-hand models are available. The Fashion mirror, Model BM-300 H (L or R), is mounted to the bike's

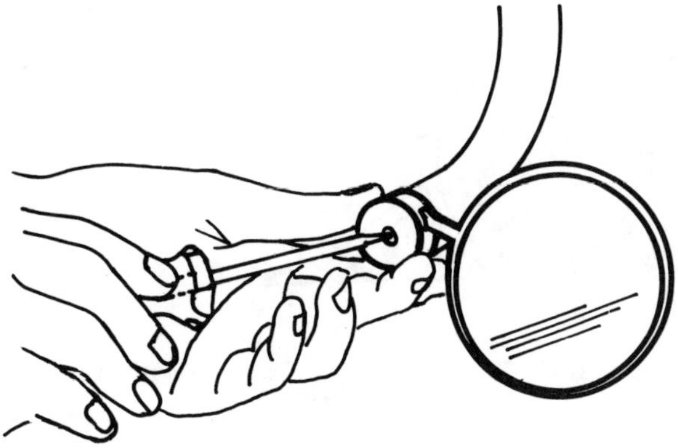

Accessories

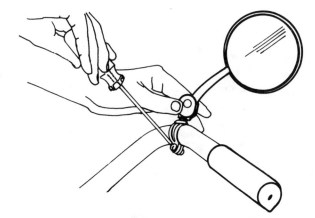

Fig. 6-11. Cat Eye's Model BM-300G, a racing bike rearview mirror.

Fig. 6-12. Cat Eye's Fashion Model BM-300H (L,R) rearview mirror.

brake lever. Model BM-300 H2 (L or R) mounts easily onto any handlebar. All three models have convex mirrors, which provide a large, clear view of the rear. Models 300-H and 300-H2 have an adjustable universal joint, operated by a coin to alter the viewing angle (See Fig. 6-13).

Reflectors

Battery-operated tail lamps are an important adjunct to safe night driving. However, if battery or lamp failure occurs, a suitable reflector can be a lifesaver. Cat Eye produces a variety of white and red wide-angle reflectors meeting regulatory standards in the U.S., Canada,

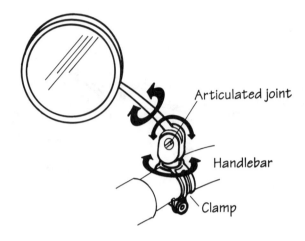

Fig. 6-13. Cat Eye's Models 300H and 300H2 rearview mirrors.

Great Britian, Australia, Belgium, France, Sweden, Germany, and Japan. Model RR-510-WU is a wheel-mounted reflector available in amber, white, or red plastic. Model RR-530 WU is only produced with a white reflector. Cat Eye's other four models are made with white reflectors for the front, red reflectors for the rear. Figure 6-14 shows Model RR-510-WU, a wheel-mounted reflector.

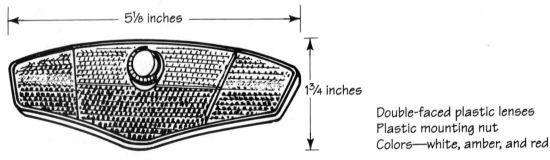

Double-faced plastic lenses
Plastic mounting nut
Colors—white, amber, and red

Fig. 6-14. Cat Eye's Model RR-510-WU wheel-mounted reflector.

Figure 6-15 shows the mounting ease of a Cat Eye pedal reflector. These pedal reflectors come in white and amber, and are manufactured according to world standards. Red rear reflectors, either mudguard or seat-stay-mounted are illustrated in Fig. 6-16.

Accessories   **89**

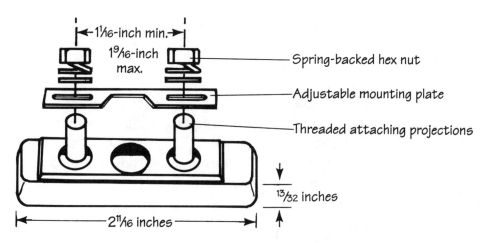

Fig. 6-15. Cat Eye's pedal reflectors.

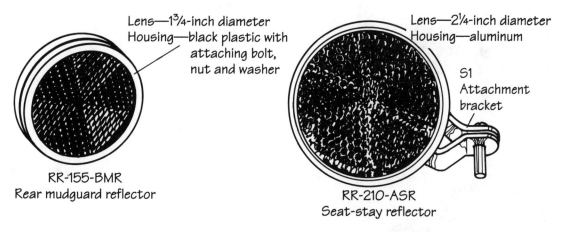

Fig. 6-16. Cat Eye's mudguard or seat-stay reflectors.

Locking Devices

A high-tech bicycle is an investment that requires protection from thieves. When you leave your bike unattended, you will need a secure method of locking wheels to the frame and the total unit to a basically immobile steel gate, fence, or lamppost.

Nevertheless, professional bike thieves now use cable-cutting pliers that can sever steel links made from 1/8 to 3/16 inches in diameter stock—or even cut through heavy aircraft cable. Hardened steel locks may offer you increased protection, but I recommend that you carry theft insurance, especially if your bike is worth more than $500.

AC International, of Sante Fe Springs, California, includes insurance coverage with the price of its locking device. It manufactures a one-piece hinged, U-shaped lock known, appropriately, as the Hinge Fig. 6-17, and includes a $5000 guarantee, no police registration required, and worldwide insurance coverage.

Fig. 6-17. AC International's Hinge bicycle-locking U.

Cobralinks (Fig. 6-18), a locking system manufactured by the J.J. Tourek Company of Elk Grove Village, Illinois, is neither a chain nor wire cable. The unit is fabricated from an interlocking series of ball-and-socket segments. The links or segments terminate with patented hardened-steel head and tail units, which are chrome-plated and key-locked. The key itself is numbered and registered only with the manufacturer, for obvious security reasons. Bike theft is practically impossible with this 3½-foot, 36-ounce "cobrahead snake." The Cobra is covered with a nylon mesh to prevent scratching the enamel on your expensive bike. This snake has neither bite nor venom, but makes the removal of your wheels and frame nearly impossible.

Fig. 6-18. J.J. Tourek's Cobralinks locking snake.

Bike Carriers

In this age of mobility, automobiles and recreational vehicles (RVs) frequently have one or more bikes attached to specialized bumper, trailer, or roof brackets. Slider Corporation of Glendale, California, has designed numerous carriers and accompanying deadbolt trailer hitch-locks for this purpose. Figure 6-19 displays a two-bike, bumper-mounted model. The deadbolt hitch-lock in Fig. 6-20 is used in conjunction with the BS-4500, a four-bike carrier.

Fig. 6-19. Slider Corp.'s two-bike car bumper-mounted carrier.

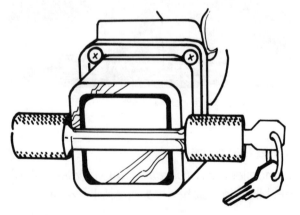

Fig. 6-20. Slider Corp.'s deadbolt hitch-lock for four-bike carrier.

Portable Working & Parking Bike Stand

Another Canadian firm, Loby's Foot America manufactures an excellent bike stand for parking, repairing, or washing your machine. This innovative workstand weighs less than 3 pounds and comes in four colors: red, white, yellow, and black. It is perfect for the cyclist who lives in a condo. Two models are available. One holds two bikes; the other, three. Loby's Foot makes handling and storage easy. It also prevents your bike from being accidentally dropped or scratched. Many cycling teams, race organizations, and competitors involved in triathlons, biathlons, and plain cycling races are enthusiastic about these fiberglass and steel lightweight, durable stands (Fig. 6-21).

Fig. 6-21. Slider Corp.'s Slider Stand, a storage or bicycle work stand.

Clothing

I have emphasized the importance of wearing approved safety helmets. Equally important are proper shoes, gloves, shorts, and shirts. Aside from the safety factor, if you belong to an active and progressive bike club, you would not wish to dress like a slob and suffer personal embarrassment. After all, gung-ho tennis players and golf enthusiasts all spend a bundle on proper equipment and clothing. Why should serious bikers be less stylish?

If you are involved in all-terrain biking, you will need a stiff, efficient shoe for the biking portion, but a comfortable one for climbing a cliff with your bike slung over your shoulder. So where do you find appropriate shoes? I recommend Specialized of Morgan Hill, California.

Specialized also offers bikers a variety of functional, fashionable shorts. Some are loose on the legs and have a cut-off point well above the knee. Others are snug, offering no wind resistance, but with less efficiency in maintaining body heat. Colors run rampant. You can find lightweight, zippered, monogrammed jackets with matching tight-fitting, below-the-knee shorts. Silk-screened, designer short-sleeved shirts, and, of course, gloves—specialized in color and function—should compliment your cycling wardrobe. Go ahead—you owe it to yourself!

On the subject of gloves—looks are important, but function is prime. The Specialized gloves are manufactured with these options: thick or minimal padding, full-finger or short-finger length, and off-road and on-road use.

Personal Waist & Backpacks

Specialized offers a full line of colorful, tough fabric packs, of varying capacity, for carrying small, personal items—even a water bottle, or a high-energy snack bar. Backpacks, of course, are larger.

Fabric panniers are slung in dual fashion, one on either side of a metal rack bolted above the rear wheel. These racks are made from an aluminum alloy to conserve weight. Dual panniers are indispensable when overnight camping is planned.

If you are a club member and your organization is anxious to promote spectator visibility, Specialized will produce team backpacks according to your club's personal specifications. The company's own Stumpjumper team uses backpacks as an advertising medium, and the team does win medals, left and right.

Water Containers & Cages

Professional road racers—and many riders who never aspire to that level of cycling—carry plastic bottles of water or fruit juice when they zip along the highways of North America or Europe. Both water containers and cages are molded from high-impact, flexible, shock-resistant, and durable materials. Cat Eye produces these bottles and cages in five distinct colors, coordinated with toe clips and straps. See Fig. 6-22 for Model BC-100.

Material—high-impact, super-tough, flexible, shock-resistant, durable plastic
Weight—39 grams
Colors—black, red, yellow, sky-blue, and white

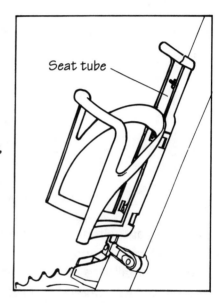

Fig. 6-22. Cat Eye's Model BC-100, a plastic water-bottle cage.

Specialized also produces water bottles and matching cages. Its prime concern is the bottle's spout design, called Many Squeeze. Apparently, some spout designs by other firms make it difficult for water from the plastic bottle to get into the rider's mouth.

Toe Clips & Straps

Cat Eye also makes toe clips, Model TC-200, and toe straps, Model TS-200 (Fig. 6-23). The clips are molded from high-impact plastic; and the straps, from durable, pliable, and water-resistant fiber.

Tools

Even though you might feel that you do not have the mechanical ability to do even minuscule, spur-of-the-moment repairs, you might one

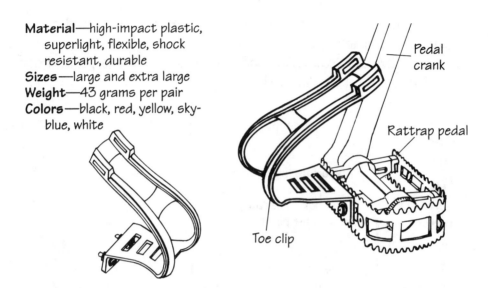

Fig. 6-23. Cat Eye's Model TS-200—plastic toe clips.

day find yourself in an emergency situation on the road with no professional repair shops nearby. With this in mind, AC International markets small kits of necessary tools including adjustable wrenches and socket wrenches (Fig. 6-24).

Fig. 6-24. AC International's portable mini-tool kit.

Rim Jimmy

Rim Jimmys (Fig. 6-25) are indestructible, nylon pocket levers, angled for easy use, which snap together to conserve storage space, namely in your pocket. An exceptionally neat spoke-tightener is illustrated in Fig. 6-26. AC International carries a wide variety of accessories, including Mr. Tuffy tire liners, which protect tubes from rim cuts, small nail holes, thorns, glass punctures, and other air-loss problems (Fig. 6-27).

Cool-Tool

A multipurpose bike repair tool, weighing less than 8 ounces, and having a folded length of only 5 inches, is now available from Cool-Tool of Chico, California. This extremely versatile, patented tool is made from S.A.E. 4130 chrome-vanadium steel, which has high tensile strength, necessary in tough loosening and tightening situations.

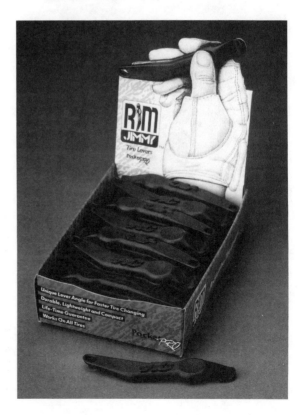

Fig. 6-25. AC International's Rim Jimmy tire lever.

Fig. 6-26. AC International's spoke tightener.

Fig. 6-27. AC International's Mr. Tuffy tire liner.

A host of different tools are included in Cool-Tool's "Friendly Get-Me-Home Companion" kit. The kit sells for $25 and provides a crescent wrench, a chain maintenance tool, three Allen wrenches (4, 5, and 6 mm), a crank-bolt socket wrench, and a Phillips screwdriver.

New to the Cool-Tool repertoire is a reversible 14/15 mm socket wrench and attachments for headset wrenches.

Bob Seals, the founder of the coolest bike tools to emerge from California in decades, has even designed a diminutive Cool-Tool saddle bag to store these tools. It is rumored that a lighter and even stronger titanium Cool-Tool is on the way. Bikers are tough, demanding people! My seal of approval is certainly on Cool-Tool. Figure 6-28 illustrates Cool-Tool in its folded position and in the chain-maintenance position.

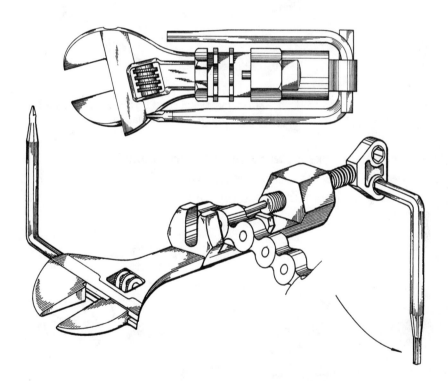

Fig. 6-28. Cool Tool, folded and chain-maintenance position.

Tube Repair Kits

Remac, Inc., of Northvale, New Jersey, markets a variety of patching materials, which include cold vulcanizing feather-edge patches; a liquid rubber buffer for cleaning the tube, prior to proper bonding of the patch; and a cold vulcanizing liquid (cement). Remac's repair kits include specific patching tools such as a stitcher, a hand buffer, buffer replacement blades, an air inflator gauge, and an air hose. Figure 6-29 illustrates Remac's products.

Fig. 6-29. Repair accessories from Remac, Inc. L–R: Perma Coil hose, corrugated stitcher, inflator gauge, straight steel hand buffer, and replacement blades.

Two mountain bike tube repair kits are shown in Figs. 6-30 and 6-31. The kit in Fig. 6-30 is useful when the puncture is not too severe, and the wheel tube is losing only a small amount of air.

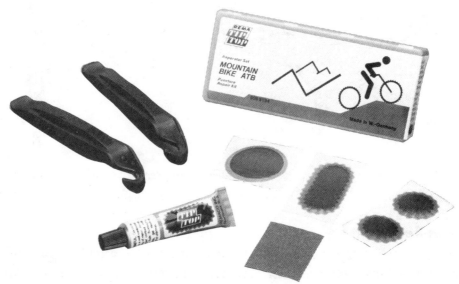

Fig. 6-30. Remac, Inc.'s mountain-bike tube repair kit.

When a total flat is apparent, you are far away from home, and a pressurized air supply is not available, ATB Kit No. 23 (Fig. 6-31) is what you need. This kit includes two 2-gas cartridges. Both kits include a tire patch, as well as numerous feather edged tube patches in different shapes. Also included are two tire levers, cold vulcanizing fluid (cement), valve adaptor components, and emery cloth. Attention, bike mountaineers! Don't leave home without these kits!

Fig. 6-31. Remac, Inc.'s mountain-bike tube repair kit, with gas cartridges.

Lubricants

Before reading about lubricants, recall my comments in chapter 4 concerning zero lubrication to moving components and the pitfalls relative to over-lubrication, not the least being the damage to clean clothes and the waste of expensive, specially formulated greases. Super Lube of Bohemia, New York, produces a wide variety of 100 percent synthetic-based lubricants, which are nontoxic and use Teflon as an additive.

This friendly-to-the-environment lubricant is

- Nontoxic
- Unaffected by temperature extremes (−65°F to +650°F)
- Impervious to oxidation
- Resistant to emulsification
- Biodegradable
- Stainless, and harmless to fabric, leather, rubber, plastic, wood and painted surfaces.

This truly remarkable lubricant is available in tubes or aerosol cans and includes a precision oiler, if needed. Figure 6-32 shows some of Super Lube's product lines.

Fig. 6-32. Super Lube's product line of lubricants.

A strong competitor to Super Lube is Finish Line Technologies, Inc., of Islip, New York. This company also produces synthetic oils and greases that are impregnated with micron-sized particles of Teflon. A quarter-ounce tube of Century, a high-viscosity synthetic lubricant has a suggested retail price of 99 cents. Finish Line also manufactures biodegradable citrus degreasers, professional polishing and protectant agents, and a super-duper tire sealant that provides permanent seals against punctures as large as 3/16-inch diameter. This sealant technology was developed in Germany. The sealant can be used on any pneumatic tire—tube or tubeless variety. No glue or latex is involved. It never dries out or freezes. At a suggested cost of $7.99 for an 8-ounce dispenser, this miracle substance will repair two tires.

Many top racing teams in the States and Canada use Finish Line products. The Coors-Light, 7-Eleven, Canadian National, and Motorola teams are but a few. Figure 6-33 shows some Finish Line products mentioned.

Fig. 6-33. Finish Line's product line of lubricants.

Chain Conditioners & Cleaners

When lubricants are used in conjunction with a well-formulated chain cleaner, your derailleur-equipped machine—especially if it is a mountain-bike—should have a transmission system that will operate as smoothly as a top.

Orleander, U.S.A., Inc., of Van Nuys, California, markets the Vetta Chain Cleaner, Model III. This compact unit is the third generation of Vetta's original chain cleaner. It's simple and efficient. As you pedal, the chain is drawn through an intricate gauntlet of specially shaped or contoured scrubbers, which quickly remove caked-up dirt, mud, and grease. This high-efficiency cleaner has no brushes to wear out or replace and will fit all derailleur bikes.

A conditioner for this mechanical cleaner is also available. It is made from natural orange rinds—even smells good, is biodegradable and environmentally harmless, and will not damage alloys, anodized metals, paint finishes, or plastics. Not only does it clean, it prevents oxidation and can be safely flushed down any sewer. Available in 16-ounce bottles, the Vetta citrus formula conditioner contains natural silicons and was chemically designed for a cleaner chain and a cleaner world. Figure 6-34 shows this third-generation scrubber-cleaner.

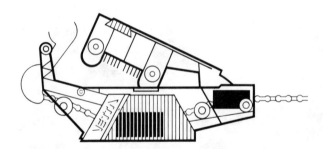

Fig. 6-34. Orleander USA, Inc.'s Vetta Chain Cleaner, Model III.

Summary

- Your bike is a moving vehicle. Be aware of state, provincial, and municipal ordinances regarding the mandatory use of lights during night and inclement weather driving, and the use of a rearview mirror and reflectors.
- Always wear an approved safety helmet.
- Bike shoes and gloves will improve your driving and prevent scraped hands in a potential spill.
- A bike computer or cyclocomputer will add interest to your biking and develop your cadence frequency, thus increasing your average bike speed. Don't forget to remove it when your bike is unattended.

- Your bike represents a considerable monetary investment. Use an approved locking system when it is not in use. Carry ample theft insurance.
- Car, recreational vehicles, and trailer carriers can transport your bike safely and simply.
- By wearing functional and attractive clothing, you are making a statement about biking. Make it an important statement. Biking is fun and healthy, regardless of one's age. (At Lakeside Village in Hobe Sound, Florida, 90-year-old Mabel Insley makes a fashion statement as she drives her three-wheeler almost every day.)
- On a long trip, especially in warm climates, carry a water bottle. Your personal waist or backpack will hold a high-energy snack, some tissues, a small first-aid kit, and some loose change, in case an emergency telephone call is required.
- A small tool kit and an emergency repair kit will make you a popular member in your club, especially if the group leader forgets his or hers.
- Use recommended lubricants for oiling and greasing moving parts. Superlube or Finish Line lubricants are both first-rate.

CHAPTER SEVEN

Safe driving techniques

A person who operates a bicycle on municipal streets must obey laws designed for all moving vehicles; no special laws apply to bicycles alone. People who disregard traffic laws or drive their bikes recklessly can be apprehended by police offers, and subsequently fined in a court of law, or they might be seriously injured. Highway driving, if permitted, might be more dangerous than city driving because of the high speeds involved.

As a responsible parent, how do you reinforce the type of thinking that turns children into responsible bicycle riders? Before I address this topic, let's talk about the stage before children get psyched on bikes.

Carriers

Small Child Carriers

Some bicycling parents transport their children on a special carrier seat that is secured to the bike frame immediately behind the seat post. The youngster is securely strapped in place and wears a proper child-size helmet. But is the little one really safe? The child's safety—or lack of it—lies in the amount of care or caution taken by the adult driver. I personally believe that any risk with youngsters is not worth taking. Is there an alternative? Leave the child at home. Obtain a babysitter. Have a relative look after your youngster. Maybe none of those options are possible, but try to find a better way. Figure 7-1 shows a typical carrier seat.

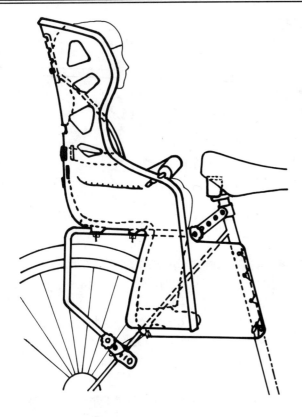

Fig. 7-1. A small child carrier.

Towed Carriers

Infants can be towed in a small, weatherproof, wheeled carrier. This device is attached to a swivel secured to the rear of the frame. A degree of left-to-right movement is obtainable. These trailers are lightweight. They do offer excellent inclement weather protection. In terms of child safety, I am not aware of statistical evidence involving accidents with attached trailers, but I strongly suspect that some exist. Again, be careful! Figure 7-2 illustrates a folded and unfolded infant transporter.

Training Wheels

When you buy your youngster a first bike, consider attaching training or balancing wheels to avoid knee and leg abrasions. These wheels act like outriggers on a catamaran. Some kids' bikes are sold as a package that includes training wheels. These wheels also can be purchased for a reasonable amount, approximately $20, and attached to the

Safe driving techniques 109

Fig. 7-2.
Top: A towed trailer for infants.
Bottom: Towed trailer in folded position.

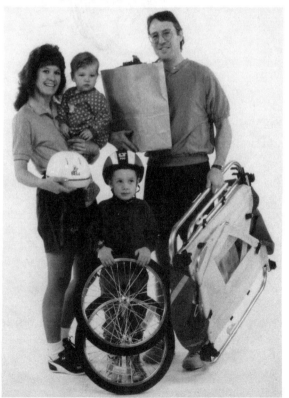

threaded shaft of the rear-wheel axle by extra hexagonal nuts and lock washers. Sheet-metal angular stays or brackets are also attached to the assembly, thereby achieving the necessary rigidity. An instructional sheet accompanying the wheel package will give ample directions for proper installation (Figs. 7-3, 7-4, and 7-5).

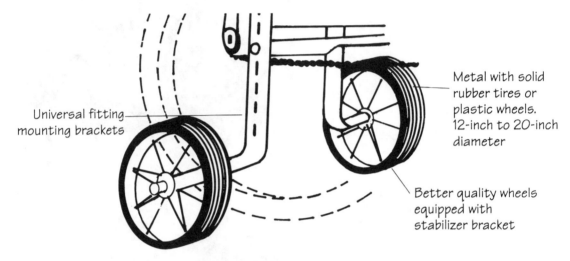

Fig. 7-3. Training wheels.

Fig. 7-4. Training wheels attached to child's single-speed, coaster-brake bike.

Fig. 7-5. The author's grandson, Adam Renton, of Brandon, Manitoba, Canada, on a Canadian Tire Corporation's children's Mightcat Supercycle, with training wheels.

Take your child to a playground or paved schoolyard. The area should be devoid of moving cars and motorcycles, and of other youngsters. Depending on the aggressiveness and fearlessness of your child, a lot of encouragement or a minimum amount might be required, before the child is "airborne." And I don't mean that in the literal sense! But even that is bound to happen on occasion. Once again, I stress the need for your child to wear a properly fitting, approved safety helmet. I know that some professional bike club members are not gung-ho on helmet wearing. I am! In a number of documented cases, people have survived serious accidents, not because of good luck, but because they were wearing approved bicycle safety helmets.

Kids & Safety Rules

The first time your child rides a bicycle, with or without training wheels, you should impress upon your youngster the need to ride according to safe biking practices. Even a toddler on a tricycle (Fig. 7-6) should be taught concepts of courtesy not to hog the sidewalk. Before your child graduates to a two-wheeler (Fig. 7-7), stress the need to drive carefully and safely. Remember that each child is different and should be guided and advised on an individual basis.

Fig. 7-6. Three-and-a-half-year-old Curtis Renton, the author's grandson, is all smiles on his trike.

I vividly recall an accident I had some 55 years ago, when I was an *Ottawa Journal* newspaper carrier. I returned home by bike after finishing my route, but with an extra paper, undoubtedly because I had missed a valued customer. I was paid nine cents per week, per customer. The phone rang, an irate person complained, and I was on my way, hellbent to deliver the paper promptly. Afterward, I was anxious to return home. My pedal cranks were revolving at a furious rate. I totally ignored a four-way intersection with four STOP signs. Wham! I hit a car at a 90-degree angle to the vehicle side. A real broadside! Over the car's hood, I vaulted. I landed on all fours—two arms were

Safe driving techniques 113

Fig. 7-7. Paul Renton of Brandon, Manitoba, Canada, has advanced to the second stage of bicycling.

outstretched, and two legs made contact with the concrete sidewalk in a matter of seconds. I told my parents that my bike was ok. In fact, it was a total write-off. My guardian angel was certainly looking after me. Nevertheless, I still carry a battle wound in the form of a two-inch scar on my left knee—scarred because I never saw a doctor, nor received surgical stitches to close the gash. My mother spread some Ozinal ointment on the good-sized wound and securely bandaged the knee to stem the bloodflow. Kids were tough in the 1930s. Dollars were few, whether American or Canadian, and government health insurance plans were nonexistent.

Insurance

A wise person carries a reasonable amount of fire, theft, liability and life insurance. Talk this subject over with an insurance agent or consultant.

Even if you have all three types of insurance mentioned, talk to your agent. Ascertain your own protection and liability to the public, if your bike is involved in a traffic accident, and a second or third party is injured. Traffic accidents often involve numerous people.

I have stressed the importance of locks in chapter 6. Still, it would be wise to consult your agent on this matter—especially if you own an expensive bike. Do you need to have it locked or chained to an immovable structure when it is not being used? Your contract might have small print that voids your insurance if your bike is stolen because it was left unlocked. You won't be psyched on bikes any longer if your $1100 mountain machine "flies the coop."

Municipal Ordinances

In my hometown of Ottawa, Canada, 15 police officers patrol on bicycles, working day and night shifts from 7:00 A.M. through 3:00 A.M. Their purpose is to help with downtown traffic violations, including some pesky bikers who fail to observe and obey municipal traffic laws (ordinances). The cycling officers have even nailed impaired car drivers and drug traffickers. This bike patrol wears approved safety helmets, with POLICE suitably enscribed in bold white letters on a blue background.

Check with your local police or sheriff's department regarding any unique traffic regulations that might exist in your community. If you are stopped by a police officer while you are on your bike, and you are not aware of any traffic infraction you've made, keep your cool. Never bad-mouth the officer.

State (USA) & Provincial (Canada) Regulations

Since a bicycle is recognized as a legal vehicle in all 50 U.S. states and throughout Canada, I won't publish traffic regulations that are identical to those which drivers of automobiles, motorcycles, and mopeds must obey. These regulations can be obtained for the asking from licensing bureaus and other state or provincial governmental agencies.

However, on some busy highways, bicycles are not permitted. Clearly designated signs will indicate that such is the case. This ordinance is not punitive. It is for your safety. Even on roads that permit bikers, I occasionally see reckless bikers, traveling two abreast. Since bikers tend to travel in a line that is not totally straight—especially at slow speeds—a wise auto driver takes heed of this situation when overtaking a bike. For the biker, a rule is to stay close to the highway shoulder. For the auto driver, the following is a safe procedure for passing a bicyclist: Scan your rearview mirror for vehicles about to overtake you on your left-hand side. When the coast is clear, guide your car a good four to six feet away from the biker, then safely pass and overtake.

The *Ottawa Citizen* of July 31, 1991, asked its readers to comment on traffic problems and hazards. According to staffer Carolyn Abrahams, "We need to teach everyone—cyclists, motorists, and pedestrians—how to survive together." Motorists commented that they were tired of cyclists acting like "yahoos and weaving between cars." Cyclists argued that they were tired of being cut off by impatient drivers. Pedestrians claimed to fear for their lives when stray cyclists whipped past them on the sidewalk.

Most of the 200 readers who replied to the *Citizen's* request said that biking education should start the moment a child first gets on a tricycle or two-wheeler with training wheels. Some added that bike safety belongs in the school curriculum. Since I once taught bicycle classes for elementary school kids, I firmly agree.

Bicycle safety councils are doing much to improve and promote safe biking, but education boards need to provide more support to teachers and young adult volunteers who have an interest in promoting safe biking.

The City of Ottawa now has a bicycle safety council known as Citizens for Safe Cycling. It was founded in 1985 after seven fatal cycling accidents in the Ottawa area within one year! More cities should follow Ottawa's lead.

Bicycle Paths

Many larger municipalities in both the U.S. and Canada have, or are developing, bicycle pathways. A lack of money appears to be a perennial problem in maintaining these pathways. For example, the National Capital Commission of Ottawa is developing and repairing some of its first paths, now two decades old. Although they were initially designed and designated exclusively as bike paths, they were renamed "recreational paths" some years later, when jogging, roller-skating and skateboarding became popular. It is difficult to believe that all these activities are compatible with each other.

Endless problems confront administrations as they attempt to cope with the daily increase of bicycling citizens. Ottawa's city planner and environmental researcher, Patrick Chen, says that 50 percent of the trips made by cars in the Ottawa region could be made by bikes. Until cities are ready to develop more bicycle pathways, this happy solution to the car-pollution problem is not going to occur.

Bicycle Tours & Competitions

Many charitable organizations exploit the tremendous popularity of and interest in biking as a means of generating operating funds. Biking is a wholesome sport, and the money raised is used for extremely good purposes. Major hospital organizations, heart-fund groups, and the American Cancer Society are charitable institutions using this fund-raising technique.

Other groups, however, conduct tours strictly for the fun of it. For example, the Gainesville Cycling Club in Florida operates the annual Horse Farm Hundred tour. It has nothing to do with horses, except that the terrain used is in the lush horse country of north Florida. Like similar tours throughout North America, bikers have the opportunity of experiencing the breathtaking beauty of the land in a first-hand way—not to mention fresh air and exercise.

The pinnacle of all biking tours is the Tour de France. The Tour is a tough one and attracts tough competitors—like two-time winner, Greg LeMond. LeMond was nearly killed in a hunting accident a few years ago, but his determination to win the tour inspired him to fight his way back to good health. Competitive biking can be more than a goal; it can be a lifesaver.

Practical Driving Skills

The SIPDA Formula

The Bicycle Corporation of America (B.C.A.) in Bethlehem, Pennsylvania, has published an excellent booklet, "Bicycle Driver's Guide." If you are purchasing a B.C.A. bike, ask the dealer for your copy. This 41-page booklet is well written, has first-rate black and white sketches, and covers a wide range of general biking topics. The sections covering laws and safety rules, traffic controls, and traffic situations are outstanding.

Safe driving techniques

B.C.A.'s staff has developed a formula for safe driving, known as SIPDA:

- S—Scan: Search to the rear, the front, and in all directions, before selecting the appropriate course for your bike.
- I—Identify: What is the hazardous traffic? Try and note it at a glance.
- P—Predict: Process this information. Predict the safest course.
- D—Decide: Make your decision. It'll be a trade-off between the safest and fastest route.
- A—Act: Carry out your decision in a safe, responsible, and courteous manner.

Figure 7-8 illustrates the SIPDA technique.

Fig. 7-8. Bicycle Corporation of America's SIPDA technique for safe bicycle driving.

Bicycle Skills Testing

B.C.A., in its "Driver's Guide Booklet," outlines a seven-exercise, practical testing program geared toward young, inexperienced bikers. The B.C.A. Skills Test has been designed so that a perfect score of 70 can be attained. Helpful diagrams and scoring columns are included.

A similar skill test, the "Go Safely Cycler's Course," is available from the Province of Ontario, Canada, Ministry of Transportation (M.O.T.). Its scoring system is different from B.C.A's; nor does it have the diagrams—a major drawback. A possible 150 points can be won; 100 is deemed a passing grade. Figures 7-9 and 7-10 show samples from the M.O.T.'s Go Safely Cycler's Course.

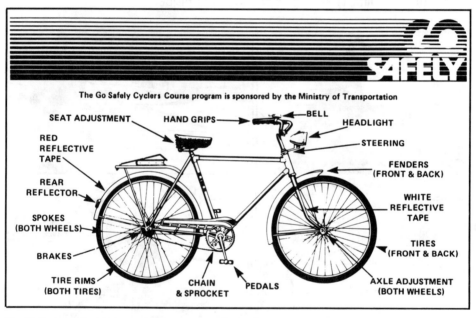

Fig. 7-9. Ontario Ministry of Transportation's Go Safely program—front side of bicycle safety inspection card.

The Ministry of Transportation test has six testing areas; the B.C.A., seven. Their testing areas are not identical, but both serve the same purpose—to ensure that young bikers are safer, more reliable, and courteous in their driving skills.

The Canadian Association of Chiefs of Police, in association with PetroCanada—a major Canadian petroleum producer—have teamed

Safe driving techniques

```
GO SAFELY CYCLERS COURSE
BICYCLE SAFETY INSPECTION: The instructor is to inspect each bike or he may
accept a card signed by an authorized bicycle dealer as proof of safe condition.

Name of Bicycle Owner _____

Inspection Made by _____
```

	Pass	Adjusted	Repairs Needed
HANDLE BARS — Right height, tight. Handlegrips must be tight.			
BRAKES MUST HOLD			
SADDLE — Should be able to reach ground comfortably with ball of foot when sitting upright on seat.			
PEDALS — Should have good treads, lubricated to turn freely.			
WHEELS — Wobble indicates need of wheel cone adjustment or replacement of broken spokes.			
REAR REFLECTOR — Required by law.			
REFLECTIVE TAPE — See requirements in Bicyclist's Handbook.			
BELL — Law requires a bell, horn or gong.			
TIRES — Should have good tread and properly inflated.			

HEADLIGHT: The program does not insist on a headlight, but the child must understand that it is required by law if riding after dark.

Fig. 7-10. Ontario Ministry of Transportation's Go Safely program—reverse side of bicycle safety inspection card.

up to produce a bicycle driver's safety program, known as Right Riders. Included with the program is a bicycle safety manual. The manual includes such advice as "Right Riders never listen to portable stereos or wear headphones when driving their bikes" and "Right Riders are always bright—especially at night."

The following five-stanza poem comes from the Right Rider's program and sums up safe driving in this lyrical fashion:

1. Always check your bike for size
   Right Riders check 'cause they are wise . . .
   Brakes, chains, tires and bells, too
   Steering, reflectors,—that's for you!
   Careful drivers always be
   Start off right—there's the key!

2. The signs are there to give clues
   For safety rules you'll always use.
   Stop signs, red, yellow lights, too
   They mean STOP—STOP for you.

Green means go, but check all ways
   Before you cross; safety pays.

3. Right Riders make sure they are seen.
   Reflectors, lights show that they're keen.
   Hand signals—left hand please—
   And driving on the right will ease
   Your trip through traffic, the law's for you
   And all other drivers, too.

4. Riding double, doing tricks
   Right Riders never do kicks
   It isn't safe; you're far too smart
   Bad habits you don't want to start
   Your helmet is a safety tool
   To protect your head and it looks cool.

5. Now you're ready, tell your mom
   Where you'll go to have your fun
   It doesn't hurt to tell her, too
   Who you're with and what you'll do.
   Right Riders follow all the rules.
   Like treasure maps, they're useful tools.

Summary

- Cannondale and other competent manufacturers produce high-quality child carriers and infant trailers. These units are only as safe as the adult biker's driving skills.
- Make sure that youngsters have training wheels for their bikes. The transition from trikes to bikes will be less painful.
- Impart safe driving concepts to your children when they are in the trike stage.
- A bicycle is a legal moving vehicle. See your insurance agent regarding any coverage you should have.
- Know your local municipal driving laws.
- Highway driving for bicycles is not permitted in some areas.
- School boards, newspapers, and municipalities are becoming more and more concerned with bicycle safety.

- The need for exclusive bicycle pathways is more pressing than ever.
- Bicycle Corporation of America produces an outstanding booklet, "The Bicycle Driver's Guide." Its SIPDA technique is great for novice bikers, as well as the older crowd.
- PetroCanada's Right Riders program—in conjunction with the Canadian Association of Chiefs of Police—instills safe driving habits in elementary school kids.
- Ontario's Ministry of Transportation's Go Safely Cycler's Course and B.C.A. offer similar practical skills testing programs.

CHAPTER EIGHT

High-tech developments

Throughout the previous seven chapters, certain components that I mentioned fall within the high-tech realm. Chapter 8 may be seen as an overview of the latest exotic materials and space-age attempts to solve problems and improve technology in biking. The following has been organized in a loose order of importance:

- Frames
    - ~ aluminum alloy
    - ~ magnesium alloy
    - ~ titanium-steel alloy
    - ~ carbon fiber

- Wheels
    - ~ Kevlar filament, tensioned
    - ~ fiberglass
    - ~ three-spoked, carbon composite
    - ~ five-spoked, injection molded

- Lights
    - ~ halogen
    - ~ light-emitting diodes

- Electronics
    - ~ cycle computers
    - ~ indicators

- Tires
    - ~ multicellular polyurethane
    - ~ Kevlar
    - ~ puncture-proof
    - ~ air-cushioned inner tube

Types of Frames

Aluminum-Alloy Frames

A number of firms produce frames from an aluminum alloy. In order to compensate for its reduced strength when compared to a steel-alloy tube of identical diameter, the aluminum tubing used is oversized and formed with an oval shape. Univega has moved one step further. It produces a patented Bi-Axial Power Oval tube that minimizes frame flex and maximizes torsional rigidity. The tubing is seamless, has a "Duralustre" finish, and is T.I.G. (thermal, inert-gas)-welded as the frame is assembled (Fig. 8-1). Easton, of Van Nuys, California, also produces high-class aluminum-alloy tubing, under the trade name of Varilite.

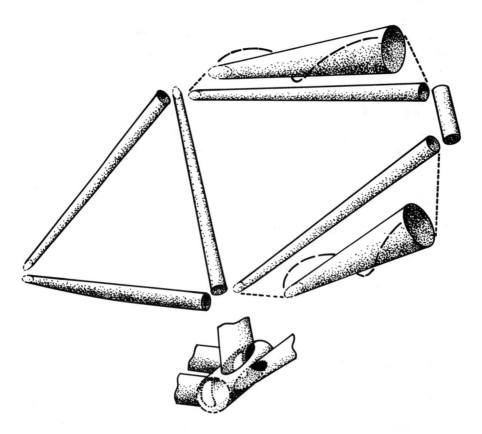

Fig. 8-1. Univega's Bi-Axial Power Oval frame tubing.

Magnesium-Alloy Frames

Magnesium die-cast frames are expertly manufactured by Kirk Bicycles USA, Inc., of Menlo Park, California. These unique mountain bike frames are marketed as either competition or touring models (Fig. 8-2). The finished bike, known as a Road Ranger, is extremely lightweight and strong. When the magnesium is alloyed with approximately 10 percent aluminum and fractional percentage amounts of zinc and manganese, the resultant alloy is noted for its high strength.

While I'm on the topic of magnesium, Time Sport produces a first-rate magnesium-alloy pedal, under the trade name Titan. What name could be more apropos when biking giants like LeMond, Fignon, and Delgoada have used these pedals in the Tour de France?

Titanium-Steel Alloy Frames

Titanium is alloyed with manganese or ferrochromium to produce a metal having these properties:

- Lightweight
- High strength
- Corrosion resistance

It weighs approximately 44 percent less than stainless and other alloy steels. Depending on the percentage of ferrochromium or manganese added to the melt—which can vary between three to seven percent—the resultant yield strength can range from 40,000 to 160,000 psi. Although the cost factor is high, lightweight and immensely strong frames for professional road racing bikes are now being made with titanium-steel alloy tubing.

America's Ken Carpenter, hailed as one of the world's five fastest riders, rides a Merlin titanium-steel alloy bike. Titanium is also used in handlebars for rugged mountain bikes. S.R. Sakae U.S.A., of Kent, Washington, markets a top-of-the-line titanium handlebar, the "Powerbulge" (Fig. 8-3).

Litespeed, the largest American producer of titanium frames and components, has 40 years of metal fabrication experience at its manufacturing outlets in Tennessee and Florida. Litespeed's products include handlebars, stems, spokes, seat posts, bottom brackets, road racer frames, and, of course, really tough mountain bike

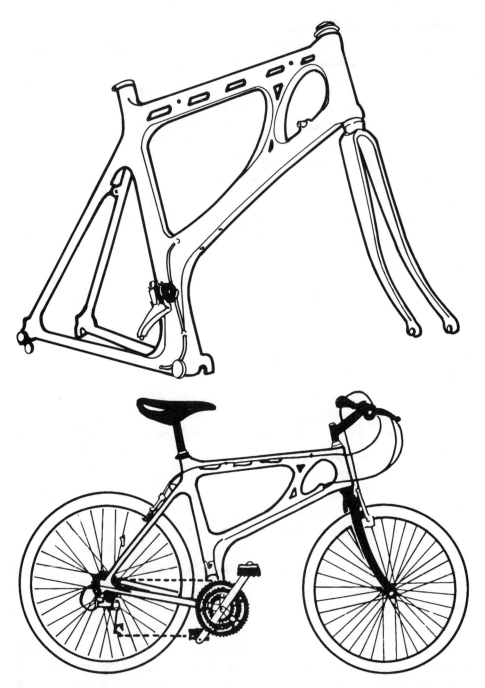

Fig. 8-2. Kirk Bicycles USA, Inc.'s precision die-cast magnesium frame; Road Ranger mountain bike.

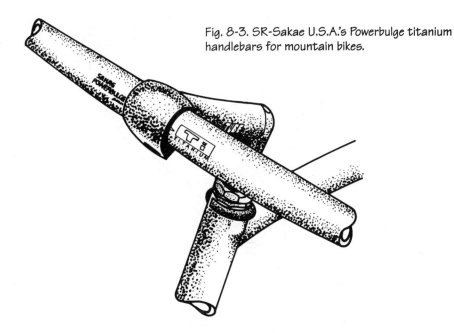

Fig. 8-3. SR-Sakae U.S.A.'s Powerbulge titanium handlebars for mountain bikes.

forks and frames. When you need a bike that is light and speedy, you can't go wrong with Litespeed.

Carbon Fiber Frames

I have been unable to obtain technical information about the manufacturing process or carbon fiber frames. Giant Manufacturing Co., Ltd., of Taiwan, makes the Giant Cadex 8901 carbon-fiber frame for all-terrain bikes and the Giant Cadex 980C frame for road-racing bikes. Basso Company of Italy manufactures the Graftek bicycle frame (Fig. 8-4). The company's U.S. importer is Security Bicycle Accessories of Hempstead, New York.

Types of Wheels

Kevlar

Kevlar, tensioned-filament or stringed type, is often used in superior wheels. Kevlar is a registered trade name of a polyaramide, synthetic material produced by Dupont. The filament or string has considerable strength under tension, and, of course, a wheel design uses this physical property. Additionally, the continuously threaded Kevlar string is

Fig. 8-4. Basso Co. Italy's Graftek bicycle frame.

sandwiched between two thin plastic disks, producing very light and rigid wheels, such as those produced by Sugino of Japan. A technical illustrator's concept is shown in chapter 2, Fig. 2-12. Note the compression disks on both sides of the plastic laminate.

Fiberglass and Injection-Molded Wheels

Alesa, a fiberglass racing disk that does not use spokes, is manufactured in Sweden. A molded depression in the fiberglass permits the biker to thread on an air pump. Figure 2-11, in chapter 2, is a technical illustrator's concept of this novel wheel.

The Aerospoke Corporation of Milford, Minnesota, manufactures a five-spoked, thin-profile, aerodynamically designed wheel from a plastic-like material. The company uses an injection-molding process. Spokes and rim are integrally formed. Because of the streamlined cross section of the spokes, and the wheel's thinness, drag and wind turbulence are reduced to a low level. This aerodynamic beauty (Fig. 8-5) retails at approximately $450 per wheel.

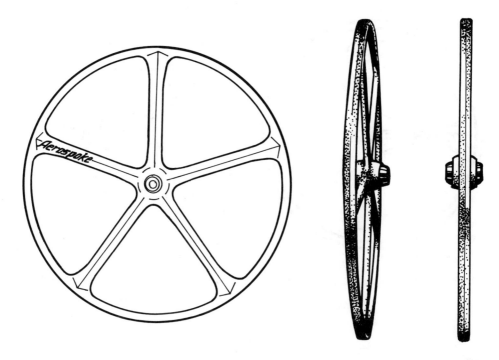

Fig. 8-5. AeroSpoke's injection-molded plastic-like material in a five-spoked, aerodynamic wheel.

Carbon-Composite Wheels

Cobalt Cycles, also known as TriSpoke Composites, Inc., has introduced a revolutionary space-age carbon-composite wheel that the company claims is the fastest wheel in the world (Fig. 8-6). If LeMond is the biker, this claim could well be true! The spokes (all three of them) are shaped like air foils. They actually generate some lift and perform even better in a breeze. At a cost of less than $500 per wheel, these 800-gram super wheels are a sure bet for USCF time trials and mass-start road-racing.

Segmented Plastic

A. Denton, a Texas plant, is now producing a 24-inch 910-gram wheel with five hollow spokes and five removable rim segments. The Polymaster 24 uses individual molds for rim segments, spokes, and hub, which permit the wheel to be repairable. A 20-inch BMX wheel is also being marketed.

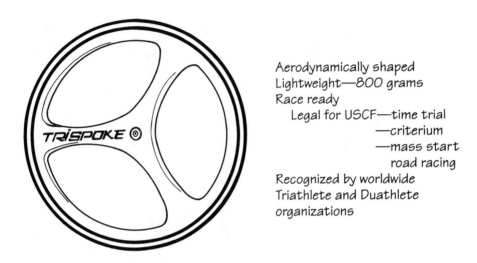

Fig. 8-6. TriSpoke's carbon-composite material in a three-spoked, aerodynamic wheel.

Lights

Halogen Lights

A number of manufacturers now produce bike-mounted halogen headlights and halogen lights that attach to the biker's safety helmet. Halogen lamps produce a stronger and more intense light than do lamps using removable incandescent bulbs. A halogen lamp is a sealed unit and is discarded when it ceases to operate. Bikers also have the option of purchasing disposable dry-cell batteries or rechargeable nicad batteries.

LEDs

Solid-state electronic components, known as LEDs (light-emitting diodes), are finding their way into bike industry electronics. When the diodes in a light unit are energized by a very low level of electrical current, they fire off a burst of light energy many times stronger than the input level needed to activate them. They are highly visible at dusk or at night—upwards to 2000 feet for a flashing red light, and 1000 feet for a flashing white light—and their electrical current consumption is minimal. Two "AA" dry cell batteries will operate a unit continuously up to 500 hours. Vistalite of Lancaster, Pennsylvania, can hardly keep up with consumer demand. Illustrations of halogens and LEDs appear in chapter 6.

Electronics

Cycle Computers

A wide variety of calculators and indicators are available. Some, like Timex's cross-training watch with an eight-lap memory, are secured to the biker's wrist.

Avocet Incorporated of California offers about 12 cycling computers. The Altimeter 50 and the Cyclometer 30 and 40 are examples.

Two Advent cyclocomputers are distributed by the Service Supply Company. The Advent AC-1000 offers current speed, average speed, trip distance, and odometer readings measured in miles or kilometers per hour. A stopwatch function and optional cadence are other available features. The Advent AC-5000 is somewhat smaller and has fewer functions, but is less expensive.

Cat Eye of Osaka, Japan, also carries a full line of excellent cyclocomputers. Some of these are discussed in chapter 6.

Vetta is also getting into the electronics field. The Vetta C-10 cyclocomputer has a molded body and comes in eight different hues. Selling for about $30, the Vetta C-10 is marketed by Orleander USA, Inc., of Van Nuys, California.

Monitors

Bikers who want to monitor their heart and pulse rates should contact the Computer Instruments Corporation of Hempstead, New York. The company's wireless monitors or indicators include the Pro Trainer, Trainer, Heart Speedometer, Heart Watch, and Deluxe Pulseminder. Heart-rate monitors and pulsemeters are the company's only products.

Tires

Three tire manufacturers should be included in this high-tech chapter.

No-Morflats

No-Morflats (Fig. 8-7) are actually special puncture-proof tubes that are placed in regular tires before they are clinched to the rim.

132 Psyched on Bikes

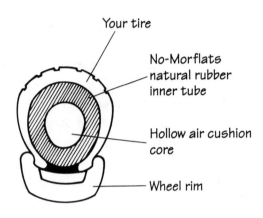

Fig. 8-7. Pictorial illustration of No-Morflats rubber/polymer tube.

Puncture-proof tube

The tubes are made from variable-density air-cushioned rubber and lightweight polymers. These specialized tubes are manufactured to fit road racers and BMX machines. The tubes will not go flat, are tough, and make riding easy. Denver, Colorado, is the home of No-Morflats.

Kevlar

As I've already mentioned, Dupont holds patents on the synthetic material, Kevlar, used in forming tensioned-string wheels and tires of advanced design and construction.

PolyAir

PolyAir of Calgary, Alberta, Canada, manufactures a unique multicellular, polyurethane, airless tire. It is not truly airless—hundreds of thousands of microscopic air cells are held within its tough matrix. Figures 2-15 and 2-16 in chapter 2 show its construction uniqueness and wheel rim application.

Summary

- Materials now used to build exceedingly lightweight but strong bike frames include the following:
    ~ Biaxial, aluminum-alloy tubing
    ~ Magnesium-alloy die-cast frames
    ~ T.I.G. (thermal, inert gas) welding to join aluminum-alloy or magnesium-alloy structures
    ~ Carbon-composite fiber tubing
- What's new and improved in wheel fabrication?
    ~ Tensioned and sandwiched Kevlar "string"
    ~ Outwards-dished, fiberglass-disked
    ~ Three-spoked, aerodynamic-shaped, carbon-composite
    ~ Five-spoked, aerodynamic-shaped, injection-molded plastic
    ~ No spokes come loose, bend or break, resulting in straight tracking without wheel wobble
- Halogen lamps produce a superior light beam. When used with rechargeable, dry-cell batteries, the system is practical and economical.
- Flashing white front reflectors and flashing red rear reflectors use light-emitting, solid-state diodes. Consuming very small amounts of electrical current, these compact units have a continuous flashing light for 200 to 500 hours, on two AA dry-cell batteries. Visibility ranges from 1000 to 2000 feet.
- All types of easy-to-read, easy-to-operate cyclocomputers are now available. Prices range from $30 to $100.
- Tired of flat or leaking tires? No-Morflats or PolyAirs are the answer. Kevlars are no slouches either but are more likely to get flats.

CHAPTER NINE

All-terrain & mountain bikes

During the past few years, all-terrain biking has become very popular. Adventurous riders want a bike that can navigate smooth fields, dirt roads, and rough landscape. When bike operators find staying upright on two wheels a difficult task, the rules of the game permit them to dismount and carry, or lug, their bikes until they can be driven under pedal power once again.

Cost

ATB clubs have stringent rules and regulations for their members. To bike, climb, and lug your bike from Point A to Point B in the shortest time is the name of the game. You have to be psyched on bikes to enjoy this strenuous pastime. Amazingly, many young and not-so-young people of both sexes are willing to cough up the fairly substantial funds required to purchase a good-quality all-terrain bike. An inexpensive ATB can cost as little as $200 on sale. One that will survive formidable usage on rough terrain can cost upwards to $1,000 or more. The price will depend on the bike's quality and the accessories included.

Manufacturers

In chapter 3, I mentioned the Montague Biframe (Fig. 3-19), a folding, full-size ATB, manufactured in Cambridge, Massachusetts. This top-performing all-terrain bicycle collapses neatly into a package that can be stored in a closet or in a car trunk.

In Rome, do as the Romans do. In Milan, do as the Regina-Ciclo bicycle company does best. Regina-Ciclo produces first-rate bicycle components and complete assemblies or groupings. This major Italian company manufactures three distinct front-wheel fork assemblies for deluxe mountain bikes (Fig. 9-1). These assemblies are priced according to their design criteria and include the standard MTB (below left), the mid-priced Professional (below middle), and the high-priced Evolution (below right).

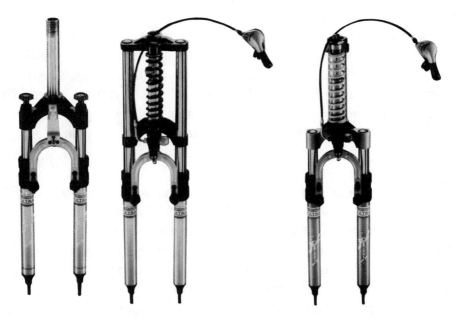

Fig. 9-1. Regina-Ciclo's MTB mountain-bike fork assemblies.

Aside from their overall weight and price differential, all components of these fork suspensions are made from chrome-moly steels. Some items are forged; others are heat-treated, hard-chromed, or formed from anticorrosive alloys. All designs have a hydraulic and spring operating system. However, the MTB model does not have a central shock absorber.

The Professional has a central shock absorber, with a substantial spring encompassing the absorber unit, as does the Evolution model. Both types have a control lever mounted on the handlebar, which permits the rider to control how much shock is absorbed during the compression phase.

Figure 9-2 provides an engineering drawing of the Regina Evolution fork; Fig. 9-3 is a photograph of the Enduro 1992 mountain bike. The MTB, the Professional, and the Evolution are all variations on the Regina Extra mountain bike fork. You will probably have to fork out a small bundle of greenbacks for an ATB or mountain bike having this superb handle. But it's worth the expense.

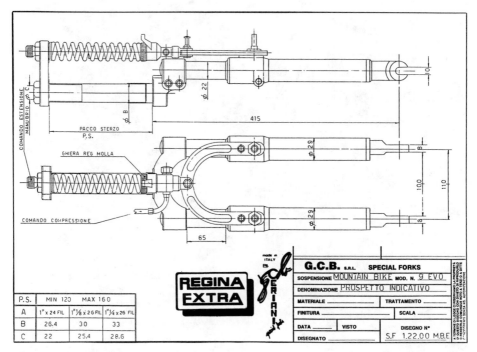

Fig. 9-2. Engineering drawing of Regina Ciclo's Evolution mountain-bike fork assembly.

For those who wish to buy American, the following ATB and mountain bikes are manufactured in the U.S.

Fig. 9-3. Regina-Ciclo's Enduro 1992 MTB.

All-terrain & mountain bikes

In Hauppauge, New York, Bike Rack Incorporated manufactures mountain, all-terrain, and cross-training bikes. Bike Rack's mountain bike has a lightweight, carbon-fiber tubing frame of high strength. It has Richey mega-bike tires guaranteed to eat up any mountain trail. The MT-1000C is, no doubt, the best and probably the most expensive bike that Bike Rack, Inc. produces, under the trade name Iron Horse (Fig. 9-4). Figure 9-5 shows an all-terrain Iron Horse. The mountain

Fig. 9-4. The Iron Horse MT-1000C mountain bike.

Fig. 9-5. The Iron Horse AT-1800 all-terrain bike.

bike has a lightweight frame made of a carbon-fiber/Kevlar-molded composite construction. The handlebars are titanium. All other components are also first-rate. The all-terrain bike has a tough, lightweight, chrome-moly frame and front-wheel fork.

The XT-5000 is the company's top-of-the-line cross-training bike (Fig. 9-6). It has a chrome-moly frame and fork, and Shimano rapid-fire gear shifters. The quick-release wheels have sealed bearings. This bike is for the committed athlete.

Fig. 9-6. The Iron Horse XT-5000 cross-training bike.

Another leader of U.S.-manufactured mountain bikes is KHS of Carson, California. This company has a worldwide distribution network in 11 countries. KHS's bikes have a 25-year warranty and are used by the John Howard School of Bicycle Champions. These two features attest to the superlative nature of KHS bikes. (Keith McLaughlin, U.S. National Champion, rides a KHS 21-speed Montana Team model.)

KHS also markets racing shorts, helmets, water bottles, and a variety of shirts. Its bike line includes 14 different models of sport, ATBs, mountain bikes, and junior models for youngsters. Figures 9-7 through 9-9 show some of the KHS product line.

Fig. 9-7. KHS's Montana Pro mountain bike.

Fig. 9-8. KHS's Montana Trail all-terrain bike.

Fig. 9-9. KHS's Cross Sport cross-training bike.

A particularly interesting and uniquely designed mountain bike is manufactured by the Cannondale Corporation of Georgetown, Connecticut. The company uses aluminum-alloy tubing in its frame construction, which reduces weight, but provides a strong and rigid frame without undue flex.

Cannondale's SE-2000 (Fig. 9-10), has a specially designed shock-mounted and pivoting rear fork. The SE-2000 is painted jet black. The swing arm, which allows the fork to pivot, is a greenish hue. This design—called the Cannondale pepperoni—is indeed hot stuff. The Connecticut Yankees sure design and fabricate hot MTBs.

Several other U.S. companies cover a share of the rapidly expanding bike market by producing components and accessories. Orleander, USA, Inc., of Van Nuys, California, manufactures its Vetta Gel brand of handlebar mountain grips (Fig. 9-11), designed to absorb shock to the operator's hands. The grip is shaped to fit the average-sized hand and uses a shaped gel inserted and bonded into an outer lycra layer to provide an extremely comfortable cushioning effect. Your pinkies never had it so good.

Fig. 9-10. Cannondale's SE-2000 aluminum alloy frame, with shock-absorbing rear fork for an MTB.

Fig. 9-11. Orleander USA, Inc.'s Vetta Gel mountain grips.

I have previously mentioned Joe Breeze Cycles of Fairfax, California, the designer and manufacturer of specialized saddles for mountain bikes. Figure 3-5 in chapter 3 shows the exclusive, auto-adjusting Hite Rite seat. Its most significant feature is that the rider is not required to dismount to adjust the seat height. Time is of the essence in winning. By not having to dismount as frequently as a competitor, your odds of winning are increased.

Summary

- All-terrain bikes and mountain bikes are popular with active cyclists in the 1990s.
- Financial investment is necessary for a quality ATB.
- U.S. manufacturing companies include Cannondale, Bike Rack, Joe Breeze Cycles, Montague, and KHS.
- Regina-Ciclo, an Italian Company, is a world leader in mountain bikes.

Appendix

Manufacturers & products

I am indebted to numerous manufacturing firms for supplying technical information and photographs for this book. The following appendix includes the names and addresses of major manufacturers or distributors and the products for which they are known. I apologize for any unintentional omissions.

AC International
11911 Hamden Pl.
Santa Fe Springs, CA 90670
Variety of accessories, including locking U mini tool kit, spoke tightener, tire levers, tire liners

Advent Service Cycle
48 Mall Drive
Commack, N.Y. 11725
Computers, helmets

Aerospoke Corporation
1200 Halden Ave.
Milford, MN 48042
Wheels

Avocet
P.O. Box 7615
Menlo Park, CA 94025
Tires, clothing, cyclometers, saddles

Barum Tires
c/o American Jawa, Ltd.
185 Express St.
Plainview, NY 11803
Tires

Bauer Cycle Supply, Inc.
404 3rd Ave.
North Minneapolis, MN 55401
Bike parts, accessories

Bellwether
11611 Mission St.
San Francisco, CA 94103
Bags

BG Products
672 Tarpon, #23
Tarpon Springs, FL 34689
Sportswear

Bianchi
385 Oyster Pt. Blvd., #6
South San Francisco, CA 94080
Road bikes

Bicycle Corporation of America
2811 Brodhead Rd.
Bethlehem, PA 18017
Bicycles

Bicycle Group
1122 Fir Ave
Blaine, WA 98230
MTBs

Bicycle Parts Pacific
P.O. Box 4750
Grand Junction, CO 81501
Power-grip pedals

Bike Lift
Maxwell Mfg.
211 E. 5th Ave., #204
Salt Lake City, UT 84103
Bike storage systems

Bike Peddler Products
5991 26th St.
Greely, CO 80634
Three-pivot-point cyclist's mirrors, bike rack

**Bike Rack USA
(Iron Horse)**
11 Constance Ct., Dept. D
Hauppauge, NY 11788
ATBs, MTBs, XTBs

Boston, Bill—Cycles
P.O. Box 114
Swedenboro, NJ 08085
Custom tandem bikes

Boyden, Tom—Fastab cycles
2706 Glenbrook Dr.
Garland, TX 75041
Custom tandem bikes

Breeze & Angell
P.O. Box 201-A
18 Meadow Way
Fairfax, CA 94930
Hite-Rite quick adjust seat locator

Bridgestone Cycle-USA, Inc.
15021 Wicks Blvd.
San Leandro, CA 94577
Top-line bike shifters

Brite Lite
P.O. Box 1386-AB99
Soquel, CA 95073
Lights

Burley Design Corp.
4080 Stewart Rd., Dept F
Eugene, OR 97402
Tandems

Campagnola-USA
43 Fairfield Pl.
West Caldwell, NJ 07006
Bike components

Canada Sachs Motor Corp., Ltd.
9615 Cote De Liesse
Dorval, P.Q., Canada H9P 1A3
Ergonomic handlebars

Manufacturers & products

Canadian Tire Corp.
Major Canadian Distributor
Head office: Toronto, Ontario
Branch stores: throughout Canada
Bikes, parts, and accessories

Cannondale Corp.
9 Brookside Pl.
Georgetown, CT 06829
MTB shock-absorbing forks, aluminum-alloy road racer

Cat Eye Co., Ltd.
2-8-25 Kuwazu Higashi-Sumiyoshi-ku
Osaka, Japan
Fitness equipment, cycle accessories, computers

Caylor Frames
P.O. Box 1793
Modesto, CA 95354
Custom tandem frames

Century Dragon Corp.
No. 2 Lane, 721 Chuncheng Rd.
Hsin Chuang City, Taipi Hsien
Taiwan
Shackle locks

Cheng Shin Rubber-USA, Inc.
545 Old Peachtree Rd.
Atlanta, GA 30174
Tires

Cinelli Cino and Co.
Primo International
10335 Landsbury, Suite 316
Houston, TX 77099
Bikes, pedals, saddles, stems

Cinetica Adjustable Handlebars
Via Francesco Vigano, N. 8-20124
Milano, Italy
Adjustable handlebars

Cobra Links Locking System
J. J. Tourek Mfg. Co.
1800-18 Touhy Ave.
Elk Grove, IL 60007
Specialized locking system

Columbus Tubing
USA Agent, I.B.P., 2131 Quail Valley East
Missouri City, TX 77459
High-tensile alloy-steel tubing

Condor Cycles (Monty Young)
90-94 Grays Inn Rd.
London, WC1X, 8AA, England
Custom bikes

Continental Roof Carriers
Graber Products
5253 Verona Rd.
Madison, WI 53711
USA only: 1-800-542-6644
Vehicle roof carriers

Cool Tool
13524 Autumn Lane
Chico, CA 95926
Compact, multifunctional bike tools

Corso Bicycle Distributors, Inc.
349 West 14th St.
N.Y., NY 10014
Bikes, bike shoes

Cuevas Custom Cycles
165 Avenue "A"
N.Y., NY 10009
Custom-made track bikes

Cycle Innovations
Sticha Ag In-der Mühle 2
CH8340 Hinwil, Switzerland
Winter MTBs

Cycle Ops
P.O. Box 1581
Healdsburg, CA 95448
Safety helmet headlights

Cycle Path
Head office: Toronto, Ontario
Franchise Br. 2515 Bank St.
Ottawa, Ontario K1V 8R9
Accessories

Cycles Peugeot-USA
555 Gotham Parkway
Carlstadt, NJ 07072
Road racing bikes

Dia-Compe, Inc.
Cane Creek Rd.
Fletcher, NC 28732
MTB brakes, brake levers

Duffy, Kevin
P.O. Box 238
Fallston, MD 21047
Rims (Van Schothurst-Rigida)

Easton Bicycle Tubing
7800 Haskell Ave.
Van Nuys, CA 91406-1999
Vanlite aluminum-alloy bike tubing

Etto Bicycle Helmets
10866 Wilshire Blvd., Suite 1270
Los Angeles, CA 90024
Safety helmets

Fastab Cycles
2706 Glenbrook Dr.
Garland, TX 75401
Custom tandem bikes

Fifth Wheel Bikerack
National sales headquarters:
11911 Hamden Pl.
Santa Fe Springs, CA 90670
Bike carriers

Finish Line Technologies, Inc.
19 Beech St.
Islip, NY 11751
Lubricants

Fuji America, Inc.
118 Bauer Dr.
Oakland, NJ 07436
Suncrest titanium ATBs

Garneau, Louis, Inc.
St. Augustin Denaures
P.Q. Canada, G3A 2E6
Clothing, safety helmets

Giant Mfg. Co., Ltd.
19 Shun Fan Road, Ta Chia Chen
Tiachung Hsien, Taiwan
ATBs, carbon-fiber frames, chrome-moly frames, MTBs

Gita Sporting Goods, Ltd. (Giordana)
Box 7266
Charlotte, NC 28241
Road bikes, clothing

Manufacturers & products

Greg LeMond Bicycles
1115 Old Bayshore Hwy.
Burlingame, CA 94011
Challenge bikes

G.T. Bicycles, Inc.
17800 Gothard St.
Huntingdon Beach, CA 92647
MTBs

G.T. Roadlight
4863 "F" Fulton Rd., N.W.
Canton, OH 44718
Lights

Haro Designs
2225 Faraday Ave., Suite A
Carlsbad, CA 92009
BMX and MTB rain jackets

Hsiang Li Industrial Co., Ltd.
323 Chunghua Rd.
Natou 54011, Taiwan
Saddles

Izumi, Pearl-USA
3630 Pearl St.
Boulder, CO 80301
Bike clothing

Jones of Boulder
P.O. Box 3096
Boulder, CO 80307
Sunglasses

Kann Mfg. Corp.
P.O. Box D-A6
Guttenburg, IA 52052
Recumbent bikes

Kashima Saddle Mfg. Co., Ltd.
30-3, 3 Chome Hannan-cho
Abeno-Ku
Osaka, Japan 545
Rain tail saddle

Kenstone Corp.
P.O. Box 132
Reynoldsburg, OH 43068
Kenda tires

KHS Bicycles
1264 E. Walnut St., Dept. 3
Carson, CA 90746
ATBs and MTBs

Kirk Bicycles-USA, Inc.
P.O. Box 866
Menlo Park, CA 94026
Die-cast magnesium frames

Klein Bicycle Corp.
2076 Prairie
Chehalis, WA 98532
Road racing bikes

Kool Stop International, Inc.
Sales Office
P.O. Box 3480
LaHabra, CA 90632
Enclosed child bike trailers

Kryptonite Locks
95 Freeport St.
Boston, MA 02122
Canada Distributor: Outdoor Gear
Montreal, Canada
Security locks

KSI Products
Cycle Components
P.O. Box 3480
LaHabra, CA 90632
Cantilever brake blocks

Laing, Colin
3454 N. 1st Ave.
Tucson, AZ 85719
Custom tandem frames, leader helmets, safety helmets

LeMond, Greg—Bicycles
1115 Old Bayshore Hwy.
Burlingame, CA 94011
Bikes, parts, accessories

Lite Speed
P.O. Box 7624
Hollywood, FL 33081
Frames, parts

Loby's Foot America
360 Notre Dame W., Suite 401
Montreal, P.Q. Canada H2Y 1T9
Portable service stands

Look Performance Sports, Inc.
Salt Lake City, UT 84103
Anatomical toe cleats

Maxam Bicycle Corp.
6691 Edwards Blvd.
Mississauga
Ontario, Canada L5T 2H8
MTBs

Marin
2066 Fourth St.
San Rafael, CA 94901
Mountain Bikes

Montague BiFrame
P.O. Box 1118
Cambridge, MA 02238
Folding MTBs

Nightsun
396 W. Washington Blvd., #600
Pasadena, CA 91103
Dual beam high-low bike headlights

No-More Flats
1438 S. Cherokee St.
Denver, CO 80223
Puncture-proof rubber and polymer tubes

Norco Products, Ltd.
Woodbridge, Ontario, Canada
Major Canadian distributor of bikes, bike parts, and bike accessories

Orleander-USA, Inc.
14553 Delano St., Suite 210
Van Nuys, CA 91411
Bicycle accessories, cyclocomputers

Overlord Industries Corp.
No. 109 Chung Shan Rd.
Shin Hua Town, Hainan Hsien, Taiwan
Bikes

Park Tool
Catalog Dept.
P.O. Box 10858
St. Paul, MN 55110-3218
Bicycle maintenance tools

Persons-Majestic Mfg. Co.
Monroeville, OH 44847
Saddles

Manufacturers & products

PolyAir Tires, Inc.
205, 259 Midpark Way, S.E.
Calgary, Alberta, Canada T2X 1M2
Semi-pneumatic tires

Profile, Inc.
6600 Armitage Ave.
Chicago, IL 60635
Airwave handlebars

Protect-A-Bike Covers
Tech Pack Cycle Components
P.O. Box 3480
LaHabra, CA 90632
Bike covers

Puch, Styr Daimler
Graz, Austria
Bikes

Raleigh (T.I.), Ltd.
22710 72nd Ave.
South Kent, WA 98032
Bikes

Regina-Ciclo S.R.L.
Via Per Meano 2
25030 Pompiano
Brescia, Italy
Drive chains, MTB forks

Remaco, Inc.
200 Paris Ave.
Northvale, NJ 07747
Tube/tire repair kits

Ross Bicycles-USA, Ltd.
P.O. Box 147
Rockaway Beach, NY 11693
Bikes

Royce Union Bicycle Co.
270 Newton Rd.
Plainview, NY 11803
Bikes

Schwinn Bicycle Co.
217 North Jefferson Ave.
Chicago, IL 60606
Bikes

Security Bicycle Accessories
P.O. Box 247
Hempstead, NY 11551
Carbon-fiber frames

Serotta Competition Bicycles
Box 106 Middle Grove Rd.
Middle Grove, NY 12850
Basso Graftek road racing bikes

Shimano American Corp.
One Shimano Dr.
Irvine, CA 92718
Transmission and brake systems

Slider Corp.
1488 Railroad St.
Glendale, CA 91204
Bike carriers

Specialized Bicycle Components
15130 Concord Circle
Morgan Hill, CA 95037
Bike accessories

S.R. Sakae-U.S.A.
18650 72nd Ave S., Dept 3
Kent, WA 98032
Titanium power bulge bars, thermally bonded FX-fork

Stitcha, Bob
Stitcha Ag In-der Mühle 2
CH8340, Hinwil, Switzerland
Winter MTBs

Sturmey-Archer of America, Inc.
1014 Caroline Dr.
West Chicago, IL 60185
Three-speed internal shift

Sun Tour
75 Digital Dr.
Novato, CA 94949
Control levers

Super Lube
24 DaVinci Dr.
Bohemia, NY 11716
Lubricants

Talegator
Graber Products
5253 Verona Rd.
Madicon, WI 53711
USA only: 1-800-542-6644
Bumper carriers

T.C. Bike Stands
1042 E. Fort Union Blvd., #383
Midvale, UT 80447
MTB and road bike stands

Thule-USA
175 Clearbrook Rd.
Elmsford, NY 10523
Roof carriers

Time Canada
P.A. Hutsebaut, Inc.
8013 Rue Alfred, Anjou, P.Q.,
Quebec, Canada H1J 1J3
Helmets, pedals, shoes

Todson, Inc. (Zéfal)
14 Connor Lane
Deer Park, NY 11729
Air gauges, air pumps

Trailmate
2359 Trailmate Dr.
Sarasota, FL 34243
Recumbents, tandem tricycles

Transpacific Sports, Inc.
23211 South Pointe Dr.
Laguna Hills, VA 92653
Fila MTBs

Trek Bicycle Corp.
801 W. Madison St.
Waterloo, WI 53594
ATBs

Trial Electric Co.
P.O. Box 45
Central Ave.
Deerfield, IL 60015
Lights

Union Fröndeburg-USA Co.
Olney, IL 62450
Lights

Velocipac
2300 Central Ave.
Boulder, CO 80301
Bags

Velo Enterprise Co., Ltd.
1012, Sec'l, Chung Shan Rd.
Tachia, Taichung, Hsien, Taiwan
Saddles

Manufacturers & products

VistaLite, Inc.
2950 Old Tree Dr., Bldg. 3
Lancaster, PA 17603
LED lights

Vittoria
Via Padre Albisetti 10
24030 Terna d'Isola (BG), Italy
Racing tires

Wheelsmith Fabrications, Inc.
3551 Haven Ave., Suite R
Menlo Park, CA 94025-1009
Spokes

Wilderness Trail Bikes
134 Redwood Ave.
Corte Madera, CA 94925
Greases

Womyn's Wheel
540 Lafayette Rd.
Hampton, NH 03841
Women's cycling products

Wonder Bike Light Electric Co.
P.O. Box 45
750 Central Ave.
Deerfield, IL 60015
Lights

Worksman Trading Corp.
94-15 100th St.
Ozone park, NY 11416
Industrial bikes

Young, Monty (Condor Cycles)
90-94 Grays Inn Rd.
London WC1X8AA, England
Custom bikes

Zéfal
Todson Incorporated
14 Connor Lane
Deer Park, NY 11729
Air gauges, air pumps

Glossary

A glossary is a collection of terms limited to a special area of knowledge or use. The technical terminology associated with bikes seems endless. I have limited this glossary to commonly used terms and definitions.

aerodynamic A scientific term relating to the streamlining of objects moving through the air, thus reducing wind resistance or drag, and permitting the moving object to travel faster with the same amount of energy used.

alkaline battery A type of nonrechargeable, expendable, direct current dry cells, using alkaline salts.

alloy The addition of various elements to aluminum or steel during the melting process, to improve ductility, hardness, and noncorrosiveness.

aluminum A lightweight element produced from clay, aluminum oxide, or bauxite by an electrolytic process.

anodized An electrolytic process of depositing a thin film of another element on aluminum or other materials, largely for appearance.

battery
- *alkaline*—a grouping of alkaline cells to produce a direct current greater than the voltage and current output of a single cell.
- *dry cell*—a general term used for all batteries not containing a liquid electrolyte; for example, automobile batteries.
- *nickel cadmium*—A type of rechargeable dry cell battery that contains nickel cadmium.
- *rechargeable*—A general term assigned to all batteries that can be recharged regardless of the chemical content.

bearing
- ball—An anti-friction device, consisting of a raceway and captive steel-alloy spherical bearings.
- needle—A small diameter, cylindrical bearing, used when bearing loads are not intensive.
- roller—A large-diameter cylindrical bearing used for heavy bearing loads.

bicycle A general term applied to all two-wheeled human-powered machines used for land transportation.

brake
- caliper—A device consisting of two plates lined with a frictional material that press against the sides of a rotating disk in certain brake systems.
- cantilever—A type of brake where both of the calipers have an arm that is pivoted about the front wheel fork.
- center-pull—A hand-operated caliper brake, having calipers on either side of the wheel rim, which are operated by a cable/linkage system that pulls centrally on both calipers when operated.
- side-pull—A caliper braking system in which the cable/linkage system operates to one side of wheel rim, but brings both calipers and their brake pads to engage the wheel rim when operated.

brazed A relatively low-temperature joining process where the steel tubes to be joined are held together by a brass alloy material. The temperature generated by the torch is not high enough to melt and cause the tubing ends to flow into each other.

butted A general technical term, indicating the place where tubes adjoin each other.

captain The lead rider of a tandem bicycle.

carbon
- filament—A type of high-tech material, where filaments or thin strands of carbon are introduced into the "mother" material to form an extremely strong and lightweight new material.
- steel—The introduction of 0.08 to 1.03 percent of carbon plus small percentages of manganese, phosphorus, and sulphur into the element iron to produce carbon steel.

carrier A general term which can describe an attachment to automobiles or RVs to carry bicycles, or attachments to bicycles to carry goods, baskets, panniers or even a child.

chafing The action of two surfaces rubbing or abrading against each other.

chain
- *drive*—The mechanism connecting the chain wheel or drive sprocket to the rear wheel or driven sprocket.
- *link*—An individual unit in a drive chain.
- *stay*—The tubular steel alloy parts of the frame, terminating in the rear-wheel dropouts. The stays act as an integral part of the bottom of the frame's parallelogram structure.
- *wheel*—The large drive sprocket, rotated by foot action against the pedal cranks.

chrome-moly steel A steel alloy containing precise amounts of chromium and moly to produce steel of high tensile strength.

cranks
- *crankset*—A general term applied to the mechanism assembly that drives the chain wheel or drive sprocket.
- *one piece*—A relatively inexpensive crank connected as a single unit to the crank hub weldment, minus, of course, the pedals.
- *pedal*—The assembly attached to the bottom of the cranks, which is free to rotate upon the application of foot and leg power.
- *removable*—cranks which can be disattached from the shaft/bearing assembly in the crank hub.

derailleur An external gear-shifting mechanism from French designers and language, literally meaning "to derail." A 3-sprocket chain wheel and a 7-sprocket free wheel are currently popular and provide 21 different speeds.

disk
- *hydraulic*—A type of disk brake in which the disk is moved against the rotating wheel by the action of hydraulic pressure of a cable-operated piston.
- *mechanical*—A brake similar to a hydraulic brake, in which the brake disk is moved against the rotating wheel disk by a hand-oper-

ated cable system. Disk brakes have better stopping power in wet weather than caliper brakes.
- *drum*—An internally expanding brake, somewhat heavy and cumbersome, manufactured primarily for children's bikes.

dropout The technical name for that portion of the front-wheel fork or rear-wheel fork that is slotted and permits the wheel assembly to be removed or dropped out quickly.

Duralumin A trade name for duraluminum, an aluminum alloy.

fender A light metal stamping of steel or aluminum alloy, or a fiberglass molded shield, covering at approximately a 90-degree angle the front wheel assembly or an 180-degree angle the rear wheel assembly, to prevent splashing of water or mud on the biker. The fenders are attached to the frame, but have braces for additional rigidity.

fork Tubular portions of the frame through which the wheel axles are held in position.

formed A technical term relating to the shaping of flat, thin metal into a three-dimensional shape, such as a wheel rim or fender.

frame The main parallelogram-shaped, tubular part of a bicycle, to which is attached the chainwheel, pedal cranks, front-wheel fork, wheel assembly, and the other necessary components that permit the frame to function as a complete bike assembly.

galvanized Steel that has been dipped in hot zinc to prevent corrosion.

generator, d.c. A compact, direct-current generating device, attached to the bicycle frame, that produces electricity by the action of a small wheel attached to the generator rotor, which is acted upon by the wheel's rim rotation.

high-tech A term to describe the latest and most exotic materials and concepts used in the manufacturing industries.

horn
- *air bulb*—A horn that gives a strong, audible signal when a flexible rubber bulb is compressed and released, thereby causing a sounding reed to vibrate.
- *Freon*—Similar to an air bulb, except that the sounding reed

vibrates under the pressure of Freon gas, contained in a small, pressurized steel bottle.

hub The enlarged cylindrical part in which the components for a front-wheel axle or a single-speed coaster brake or a three-speed internal shift or a derailleur drive mechanism are contained.

lever
- *brake*—A hand-operated pivoting arm that is cable-connected to a brake caliper.
- *shift*—A hand-operated pivoting arm that is cable-connected to a bike's gear-shifting mechanism.

link An individual unit in a chain assembly through which the sprocket teeth are engaged.

nickel-cadmium a chemical compound used in rechargeable dry-cell batteries.

pannier A general term referring to bags of different sizes, shapes, and materials, which are slung to a carrier over the rear wheel and fender, one bag on each side of the wheel.

pin, taper A hardened, tapered, steel pin, used to secure machined parts together, while permitting the two components to be disattached when the pin is removed by driving it out with a steel drift. Cylindrical pins and hollow, slotted, hardened steel pins are also used to join machined parts. An adaptation of the taper pin has a threaded small end. After the pin is driven home, a lock washer and hexagon nut are threaded in place.

post
- *saddle or seat*—A tubular steel alloy assembly, to which the saddle is bolted. The post fits snugly into the top bracket, where it is adjusted for the individual biker's height and held in place with a bolt, nut, and lock washer.
- *steering (stem)*—A similar device, but referred to as a gooseneck or stem, held in place in the headtube of the front wheel fork by an expansion bolt which threads into a tapered plug.

rechargeable battery A battery made up of a number of rechargeable dry cells, such as nickel-cadmium.

recumbent bicycle A type of machine on which a biker sits with his legs placed outwards, forming an approximately 90-degree angle to his upper body.

rim   The sheet-metal formed part of the wheel assembly, to which the spokes are fastened and are, in turn, secured to the wheel hub.

saddle or seat   A triangular shaped seat unit having built-in shock-absorbing springs. The saddle is attached to the seat post and is adjustable upwards and downwards to suit the individual biker.

screwdriver   A hand-operated tool, frequently used in bicycle maintenance. The star or Phillips design and the regular design are commonly used. Ratchet screwdrivers, with a left-handed or right-handed turning direction and a fixed position, are also available.

Shimano   A major Japanese bicycle and bike components manufacturer, having worldwide distribution outlets.

spoke   Small-diameter, wirelike components made from steel that connect the hub to the wheel rim, thus combining rigidity and strength with lightness in the wheel assembly. Spokes are either round or oval. For rust protection, the ordinary steel ones may be galvanized. Better-quality spokes are made from stainless steel.

sprocket   A toothed, external gear-like device over which the drive chain engages. Sprockets are of either the drive or driven types. The drive sprocket is located at the pedal position; the driven sprocket, at the rear wheel position.

steel   A strong metal produced when precise amounts of carbon are added to iron which are then melted together at a high temperature. The addition of specific amounts of chromium-molybdenum, nickel-chromium, titanium, or chromium-vanadium, for example, produce alloy steels with specific physical characteristics such as strength and machinability.

stem
- handlebar stems or goosenecks—fabricated in different lengths in order to vary the distance between the saddle and handlebars. The stem is held in the front wheel fork tube by a stem-expander bolt and a tapered or wedge-shaped, threaded plug.

stoker   The person located in the second position behind the captain on a tandem bike.

Sturmey-Archer A British manufacturer of a popular three-speed, internal gear shifting mechanism.

tandem A bicycle manufactured for two people.

tire
- *clincher*—Practically obsolete, at least in North America. It has a rubber bead which fits into a groove in the wheel rim.
- *tubular*—Primarily a racing tire. The tread is a separate component. The inner tube containing air is placed inside an outer fabric casing, which is sewn up.
- *wired on*—Contains a wire bead in the outer edges of the tire. The tire firmly engages the rim when the tube is pressurized.
  NOTE: The International Standardization Organization (ISO) has standardized the nomenclature for tires. Wired-on tires are now known as tires; and tubular tires, as tubulars.

tricycle A vehicle that has three, rather than two wheels. It is commonly known as a trike. One wheel is at the front and two are at the rear.

tube
- *drive*—A sophisticated engineering design, in lieu of sprockets and chain, for propelling a bicycle.
- *tire*—The envelope within the tire that holds the air under pressure.

tubing
- *double-butted*—A type of tubing used on fine bicycles. Reynolds 531 double-butted 21/24 is a prime example. The wall thickness is twice as thick, over a 3-inch butt and a 6-inch butt, at the opposing ends of a 24-inch tube. The joined yield strength of 531 D/B is 80,000 psi.
- *drawn*—Refers to the process of making seamless tubing, where the cylindrical ingot of alloy steel is drawn over a die of the proper size and shape.
- *plain gauge*—Indicates that the wall thickness is a constant throughout the length of the tube; therefore, the welded joints are considerably weaker than a double-butted tube's.
- *seamed*—A tube formed from flat stock, then welded along the joint; used in inexpensive bikes.

- **seamless**—A tube formed by the process of using solid stock drawn over specially sized and shaped materials; somewhat similar to the process of aluminum extrusion.

unicycle A single-wheeled cycle with a banana seat, used especially by acrobats and circus performers. Both Columbia and Schwinn manufacture these oddities.

valve That part of the valve assembly that protrudes from the inner surface of the tube through the wheel rim. The valve is an integral part of the tube.

- *presta*—An imported valve from Europe. The valve core is not removable. It is more susceptible to damage than a Schraeder.
- *Schraeder*—A valve core of American design. The core can be removed by inverting the valve cap and using it as a sort of screwdriver or wrench. Caps without the elongated and grooved slot at the end cannot function in this manner.

Velcro A patented fastening material used to temporarily secure fabric items together. The design was based on the natural adhering of the thistle plant to clothing.

welding
- *arc*—Synonymous with electric welding.
- *electric*—The electric arc is generated by bringing an electrode—one that carries a relatively low voltage, direct current of electricity, but with a high current flow (amperes)—in contact with the components or parts to be joined. The parts are slightly separated, which allows the current to flow through the gaseous medium, thereby creating local heat of welding intensity.
- *gas*—A welding process using oxygen and acetylene carried in two separate tanks. The amount of each gas used is varied at the flame tip, creating different temperatures.
- *oxy-actylene*—Synonymous with gas welding.
- *spot*—Another type of electric welding, often referred to as resistance welding. The sheet metal components are clamped between two electrodes, which generate sufficient heat for welding at that spot. This process is used for joining light-gauge metals in a mass-production setting.

- *T.I.G.*—Thermal inert gas is used for welding aluminum and magnesium parts. The flame tip is enveloped by inert gas, thereby preventing a harmful buildup of oxides to the weld area.

wrenches
- *Allen*—See key-Allen below.
- *closed-end*—A specific-sized wrench having a closed end, which must be placed over the bolthead or nut for adjustment.
- *crescent*—Actually a manufacturer's trade name, but long usage has made Crescent applicable to most open-ended, fixed-size wrenches.
- *hexagon*—Most wrenches are adjustable to fit hexagon-headed bolts and nuts, or capscrews.
- *key-Allen*—A hexagonal wrench, used with small machine screws to secure an internal component to a rotating shaft.
- *open-end*—Wrenches used to adjust hexagon-headed fastening devices, without placing the wrench over the head. They are frequently used where working space is limited.
- *ratchet*—A wrench having a pawl/spring mechanism that tightens and loosens fastening devices when wrench swing is limited.
- *socket*—A special type of closed-end, ratchet wrench, having sockets of different—but standard—hexagon sizes. A common wrench handle is used with the interchangeable sockets.

Index

A
A. Denton, 129
AC International, 90, 96
accessories for bikers, 2, 77-105
   backpacks, waist packs, 94
   carriers, automotive carriers, 91, 92
   chain conditioners and cleaners, 103-104
   child carriers, 107-108
   clothing, 93-94
   computers, on-board computers, 84-86, 123, 131, 133
   generators for lights, 78
   gloves, 94
   helmets, 77-78
   lights, 78-84
   locking devices, 89-91
   lubricants, 101-103
   mirrors, rearview, 86-87
   multipurpose tools, Cool Tool, 97-99
   parking stand, 93
   reflectors, 87-89
   tire levers, 97
   toe clips and straps, 95
   tools, 95-96
   training wheels, 108-111
   tube repair kits, 99-101
   water containers and cages, 94-95
Aerodynamic Competitive Edge (ACE) wheels, 20-21
Aerospoke Corporation, 128
Alesa Manufacturing Co., 16, 18
all-terrain bicycles, 135-144
   cost, 135
   features and options, 136-144
   manufacturers, 136-144
aluminum-alloy frames, 124
Archer, 56
automotive bike carriers, 91, 92
Avocet Incorporated, 131
AVRO CF-100 ultralight airplane, 47, 49

B
backpacks, 94
ball bearings, 53
Basso Company, 127
battery-powered lights, 79-84
Bayliss-Wilby, 20
bearings
   ball bearings, 53
   needle bearings, 53, 54
   roller bearings, 53, 54
Bicycle Corporation of America (BCA), 32
bicycle path etiquette, 115-116
Bike Rack Incorporated all-terrain bicycles, 139-140
Biospace chainwheel, 60-61
brake levers, 72-74
braking mechanisms, 67-76
   brake levers, 72-74
   caliper brake hybrid design, 67-68
   caliper brakes, 69-72, 76
   coaster brakes, 68-69, 75
   disk brakes, 74
   drum brakes, 74-75
   maintenance, caliper brakes, 72
   skidding, 71
Brite Lite Cycling Lights, 79, 80-81
Burley Duet and Samba, 41-42

C
caliper brake hybrid design, 67-68
caliper brakes, 69-72, 76
   maintenance, 72
   side-, center- and cantilever-pulls, 70-71
   skidding, 71
Campagnolo, 32
Cannondale, 15, 142-143
carbon fiber frames, 127
carbon-composite wheels, 129

Index

Carpenter, Ken, 125
carriers, automotive carriers, 91, 92
Cat Eye, 80, 84-86, 87, 88, 95, 131
chain conditioners and cleaners, 103-104
chain link removal, 63-64
chain removal and reinstallation, 62-63
chain tension adjustment, 62-63
child carriers, 107-108
child-safety rules, 111-113
cleaning, 56
clothing for bikers, 2, 93-94
clubs and associations for bikers, 1
coaster brakes, 68-69, 75
Cobalt Cycles, 129
collapsible bikes, 43, 136
competition events, 116
Computer Instrument Corporation, 131
computers, on-board computers, 84-86, 123, 131, 133
Cool Tools, 97-99
cost of bikes, 3, 5-7
Cycle-Shock shock-absorbing seat posts, 35, 37, 46
Czerwinksi, Waclaw, 47

D

derailleur transmissions, 59-60
designs and styles of bikes, 2-6
Desoto Classic tricycle, 33-34
disk brakes, 74
drum brakes, 74-75

E

Easton, 124
Etto Helmets, 77-78
EZ Roll Regal, 12-13, 33-34

F

fenders, 27-28
fiberglass wheels, 128
Finish Line Technologies, 102-103
frames, 9-16
    aluminum-alloy frames, 124
    carbon fiber frames, 127
    high-tech developments, 123, 124-127, 133
    magnesium-alloy frames, 125
    materials used in frame construction, 14-16
    tandem or multiseat bicycles, 11-12
    titanium-steel alloy frames, 16, 125-127
    tricycles, 12-13
    tricycles, working tricycles, 13-14
friction effects, 53

G

generators for lights, 78
Giant Manufacturing Co., 127
gloves, 94
Gossamer Albatross ultralight airplane, 47-48
greasing, 56

H

halogen lights, 130
handlebars (see steering mechanisms)
health benefits of biking, 1
helmets, 77-78
high-tech developments, 123-133
    computers, 131, 133
    electronics, 123
    frames, 123, 124-127, 133
    lights, 123, 130, 133
    monitors, 131, 133
    reflectors, 133
    tires, 123, 131-133
    wheels, 123, 127-130, 133
Hite Rite, 33, 46, 144

I

injection-molded wheels, 128-129
insurance, 113-114

J

J.J. Tourek Company, 90, 91
Joe Breeze Cycles, 33, 46, 144
John Howard School of Bicycle Champions, 140

K

Kann Manufacturing Corp., 35-36, 46
Kevlar wheels, 127-128, 132
KHS Company all-terrain bicycles, 140-142
Kirk Bicycles USA Inc., 125

L

laws and ordinances concerning biking, 114-115
le Cheminant, Arthur, 47
LED lights, 130-131
LeMond, Greg, 116
lights, 78-84
    battery-powered, 79-84
    generators, 78
    halogen lights, 130

Index

high-tech developments, 123, 130, 133
LEDs, 130-131
Litespeed, 125
Loby's Foot America, 93
locking devices, 89-91
lubrication, 55-56, 64
   chain conditioners and cleaners, 103-104
   lubricants, 101-103

M

magnesium-alloy frames, 125
maintenance, 55-56
Majestic, 20
manufacturers and products listing, 145-153
McLaughlin, Keith, 140
mirrors, 86-87
monitors, 131, 133
Montague Biframe, 43, 136
mountain bicycles (see all-terrain bicycles)
multicellular tires, 132
multiseat bicycles (see tandem bicycles)
multispeed derailleur transmission, 59-60

N

needle bearings, 53, 54
No-Morflats puncture-proof tires, 131-132

O

odometers, computerized, 84-86
oil, 55-56
one-speed transmission with coaster brake, 56-57
ordinance and laws concerning biking, 114-115

Orleander USA Inc., 104, 142-143
Ottawa Man-Powered Flight Group, 47, 49

P

parking stand, 93
pedals, 49-52
   conventional pedal assembly, 57
   materials and design, 52
   pedal crank configurations, 49-51
   toe clips and straps, 95
Penny Farthing bicycle, 47-48
PolyAir Tires Inc., 23-24, 132
products and manufacturers listing, 145-153
puncture-proof tires, 131-132

R

rearview mirrors, 86-87
recumbent bicycles
   handlebar configuration, 44
   saddles and seats, 35-36
reflectors, high-tech developments, 133
Regina-Ciclo all-terrain bicycles, 136-138
Remaco Inc., 26, 99-101
Right Riders poem, 119-120
roller bearings, 53, 54

S

S.R. Sakae USA, 125
saddles or seats, 29-37, 46
   adjustments, 29-32
   height of saddle from ground, 31

materials used in saddle construction, 32-33
   recumbent bicycles, 35-36
   shock-absorbing seat posts, 35, 37
   Trailmate saddles, 33-35
safe driving techniques, 107-121
   bicycle path etiquette, 115-116
   carriers, 107-108
   child-safety rules, 111-113
   insurance, 113-114
   ordinances and laws concerning biking, 114-115
   practical driving skills, 116-120
   SIPDA formula for safe biking, 116-117
   small-child carriers, 107-108
   testing skills, 118-119
   tours and competitions, 116
   towed carriers, 108
   training wheels, 108-111
Schwinn Bicycle Co., 12, 73, 74
Seals, Bob, 99
seats (see saddles or seats)
segmented plastic wheels, 129
semi-pneumatic tires, 23-24
Service Supply Company, 131
shift levers, 60-61
Shimano, 20, 51-53, 57-60, 67, 69, 72, 74-75, 140
SIPDA formula for safe biking, 116-117
Slider Corporation, 92

Specialized Bicycle Components Inc., 78
speedometer, computerized, 84-86
spoke wrench, 9, 10, 17
spokes, 18, 20-21
stands, parking stands, 93
steering mechanisms, 37-46
   collapsible bikes, 43
   handlebar configurations, 37-46
   safety standards, ANSI and CSA, 44
Sturmey-Archer, 1, 20, 56, 58
Sugino, 18, 128
Sun Metal Products, 16
Sun Tour USA Inc., 60-62, 72-73
Super Lube, 101-102

T
tandem bicycles
   frame construction, 11-12
   handlebar configurations, 41
testing biking skills, 118-119
three-speed internal gear-shifting mechanism, 58-59
Time Sport, 125
tire levers, 97
tires, 21-27
   air pressure recommendations, 22
   air valve positioning in rim, 25-26
   high-tech developments, 123, 131-133
   Kevlar tires, 132
   levers, Rim Jimmy tire lever, 97
   materials used in tires, 21-22
   multicellular tires, PolyAir, 132
   puncture-proof tires, No-Morflats, 131-132
   removal and replacement of tire and tube, 26-27
   repair kits, 26-27
   semi-pneumatic tires, 23-24
   tread profiles, 21-22
   tube repair kits, 99-101
   tubes, 25-27
titanium-steel alloy frames, 125-127
Todson, Inc., 16
toe clips and straps, 95
tools, 95-96
   multipurpose tools, Cool Tool, 97-99
   tire lever, Rim Jimmy, 97
Tour de France, 116
tours, 116
towed carriers, 108
Trailmate Company, 11-12, 33-35, 37, 41, 46
training wheels, 108-111
transmission systems, 47-65
   ball bearings, 53
   Biospace chainwheel, 60-61
   chain conditioners/cleaners, 103-104
   chain link removal, 63-64
   chain removal and reinstallation, 62-63
   chain tension adjustments, 62-63
   friction effects, 53
   lubrication and maintenance, 55-56, 64
   multispeed derailleurs, 59-60
   needle bearings, 53, 54
   one-speed transmission with coaster brake, 56-57
   pedal mechanisms, 49-52
   purchase tips, 64
   repairs, 64-65
   roller bearings, 53, 54
   shift levers, 60-61
   three-speed internal gear-shifting mechanism, 58-59
   ultralight aircraft using bicycle transmissions, 47
tricycles, 12-13
TriSpoke Composites Inc., 129
Tsuyama Manufacturing Co. Ltd., 86
tube repair kits, 99-101
tubes and tires, 25-27

U
Union-Frondenberg-USA, 78, 80
Univega, 124

V
Varilite, 124
Vetta, 131
Vistalite Inc., 82-84, 130-131

W
waist packs, 94
water containers and cages, 94-95
wheels, 16-21

Index

air valve positioning in rim, 25-26
carbon-composite wheels, 129
cross-sectional shapes of rims, 18
disc-shaped wheel, racing wheel, 19
fiberglass wheels, 128
high-tech developments, 123, 127-130, 133
injection-molded wheels, 128-129
Kevlar wheels, 127-128
materials used in rim construction, 15-16
removal and replacement of tire and tube, 26-27
repairing damaged rims, 17
rim alignment correction, spoke wrench use, 9, 10, 17
rim function and design, 15-18
segmented plastic wheels, 129
spokes and spoke configuration, 18, 20-21
tire installation procedure, 24
Wheelsmith, 20, 21
Wood, Phil, 74
working tricycles, 13-14
Worksman Trading Corporation, 13-14
Worksman, Morris, 13

Other Bestsellers of Related Interest

THE ILLUSTRATED VETERINARY GUIDE FOR DOGS, CATS, BIRDS, AND EXOTIC PETS—Chris C. Pinney, DVM

You'll keep your menagerie wagging, purring, chirping, hopping, or swimming with this guide. It's by far the most detailed do-it-yourself pet care manual available for dogs, cats, birds, rabbits, hamsters, and fish. You'll find sections on caring for older pets, diseases people can catch from animals, treating cancer in pets, and the difficult euthanasia decision. 704 pages, 364 illustrations. Book No. 3667, $29.95 hardcover only

GARDENING FOR A GREENER PLANET: A Chemical-Free Approach—Jonathan Erickson

Control pests in your lawn and garden with these environmentally safe methods. Using a technique known as "integrated pest management," this book shows you how to protect food and foliage from destructive insects without contamination from toxins found in chemical pesticides. He explains, in easy-to-follow steps, the correct way to use natural methods such as beneficial insects and organisms, companion planting, minerals and soaps, and botanical insecticides in the war against garden-hungry bugs. 176 pages, 108 illustrations. Book No. 3801, $13.95 paperback, $21.95 hardcover

CREATIVE GARDEN SETTINGS—John D. Webersinn and G. Daniel Keen

Look at the ways you can landscape your property and turn your house into a panorama of outdoor creativity, at the same time increasing the value of your home. Whether you want to build a deck, a patio, a stone fence, or a trickling fountain—nothing is beyond your reach. Keen and Webersinn combine their skills to bring you a well-written guide to everything from building permits to outdoor lighting. 200 pages, 100 illustrations. Book No. 3936, $14.95 paperback, $24.95 hardcover

ALTERNATIVE ENERGY PROJECTS FOR THE 1990s —John A. Kuecken

Free yourself from high energy bills by using this newly revised edition of the 50,000-copy seller. It provides clear, illustrated instructions for building instruments that can harness enough wind, water, or sunlight to meet a large percentage of home electricity needs. And it includes computer programs for making calculations, information on small windmills, and the new Power Tower. 264 pages, 140 illustrations. Book No. 3835, $14.95 paperback, $24.95 hardcover

THE NEW EXPLORER'S GUIDE TO MAPS AND COMPASSES—Percy W. Blandford

Written especially for children ages 8 and up, this is a complete guide to understanding, reading, and using maps and compasses. Filled with activities, it teaches valuable skills such as identifying latitude and longitude; reading map symbols, abbreviations, and scales; setting a map or compass and orienteering by them; planning a trip using a map; or making a map for a journey through the wilderness. 160 pages, 67 illustrations. Book No. 3859, $7.95 paperback, $15.95 hardcover

MAKING SPACE: Remodeling for More Living Area —Ernie Bryant

After you've developed your remodeling plan, this book gives you the step-by-step instructions and diagrams you need to complete your project. You'll find easy-to-follow techniques for constructing space-enhancing attic, garage, basement, full-room, and porch/deck conversions. Plus easy-to-understand instructions highlight all the important steps of construction, and lead you through the entire process. 248 pages, 262 illustrations. Book No. 3898, $12.60 paperback, $22.95 hardcover

HOW TO GET MORE MILES PER GALLON IN THE 1990s—Robert Sikorsky

This new edition of a bestseller features a wealth of commonsense tips and techniques for improving gas mileage by as much as 100 percent. Sikorsky details specific gas-saving strategies that will greatly reduce aerodynamic drag and increase engine efficiency. New to this edition is coverage of the latest fuel-conserving automotive equipment, fuel additives, engine treatments, lubricants, and maintenance procedures that can help save energy. 184 pages, 39 illustrations. Book No. 3793, $7.95 paperback, $16.95 hardcover

BRICKLAYING: A Homeowner's Illustrated Guide—Charles R. Self

In this handy do-it-yourself guide you'll learn the basics of bricklaying: how to create different pattern bonds, mix mortar, lay bricks to achieve the strongest structure, cut bricks, finish mortar joints, and estimate materials. You'll also find out how to mix, test, and pour concrete to create foundations and footings for your brickwork. With the step-by-step instructions and illustrations found here, you can build any project with little difficulty. 176 pages, 146 illustrations. Book No. 3878, $14.95 paperback, $22.95 hardcover

Prices Subject to Change Without Notice.

Look for These and Other TAB Books at Your Local Bookstore

To Order Call Toll Free 1-800-822-8158
(in PA, AK, and Canada call 717-794-2191)

or write to TAB Books, Blue Ridge Summit, PA 17294-0840.

Title	Product No.	Quantity	Price

☐ Check or money order made payable to TAB Books

Charge my ☐ VISA ☐ MasterCard ☐ American Express

Acct. No. _____ Exp. _____

Signature: _____

Name: _____

Address: _____

City: _____

State: _____ Zip: _____

Subtotal $ _____

Postage and Handling
($3.00 in U.S., $5.00 outside U.S.) $ _____

Add applicable state and local sales tax $ _____

TOTAL $ _____

TAB Books catalog free with purchase; otherwise send $1.00 in check or money order and receive $1.00 credit on your next purchase.

Orders outside U.S. must pay with international money order in U.S. dollars.

TAB Guarantee: If for any reason you are not satisfied with the book(s) you order, simply return it (them) within 15 days and receive a full refund.   BC